Stochastic Processes and Applications

Stochastic Processes and Applications

Editor

Jacek Fabian

scitus
academics

Stochastic Processes and Applications
Edited by **Jacek Fabian**

ISBN: 978-1-68117-648-2
Library of Congress Control Number: 2016946512

SCITUS Academics LLC,
616, Corporate Way, Suite 2, 4766,
Valley Cottage, NY 10989

www.scitusacademics.com

Notice

Printed in United States of America on Acid Free Paper

Preface

The field of stochastic processes is essentially a branch of probability theory, treating probabilistic models that evolve in time. It is best viewed as a branch of mathematics, starting with the axioms of probability and containing a rich and fascinating set of results following from those axioms. Although the results are applicable to many areas, they are best understood initially in terms of their mathematical structure and interrelationships. Applying axiomatic probability results to a real-world area requires creating a probability model for the given area.

Stochastic processes were first studied rigorously in the late 19th century to aid in understanding financial markets and Brownian motion. These subjects originally had an application emphasis, the first on queueing and congestion in data networks and the second on modulation and detection of signals in the presence of noise. It has become increasingly clear that the mathematical development is applicable to a much broader set of applications in engineering, operations research, physics, biology, economics, finance, statistics, etc.

Stochastic Processes and their Applications emphasizes on the theory and applications of stochastic processes. It is concerned with concepts and techniques, and is oriented towards a broad spectrum of mathematical, scientific and engineering interests. Characterization, structural properties, inference and control of stochastic processes are covered.

Table of Contents

1

CHAPTER

Valuation of Game Option Bonds under the Generalized Ho-Lee Model: A Stochastic Game Approach

Natsumi Ochiai, Masamitsu Ohnishi

Graduate School of Economics, Osaka University, Osaka, Japan

ABSTRACT

We propose a valuation for the bond in which an issuer and a holder are simultaneously granted the right to exercise a call and put options. As the term structure model of interest rate, we use the Generalized Ho-Lee model that is an arbitrage-free binomial lattice interest rate model. The issuer and the holder play a series of stage games in each exercisable node on the lattice whose payoff structure is dependent on the nodes. We formulate the valuation problem as a stochastic game or a Markov game. Our stochastic games possess saddle points in pure strategies for each stage game. We derive the optimality equation to solve backwardly the bond values and the exercise strategies from the maturity to the initial time. Our numerical results are useful to intuitively understand the risk to a change of interest rates for options embedded in bond.

Keywords: Generalized Ho-Lee Model, Callable Bond, Putable Bond, Stochastic Game, Key Rate Duration

1. INTRODUCTION

Callable and putable bonds are one of the major interest rate derivatives due to its ability to respond the needs of various markets participants. Callable bonds give an issuer the right which can call a bond before the maturity date. The issuer is allowed to exercise the right at the several prescribed exercise times. If the issuer exercises the option, she buys back the bond at a call price from a holder. Then, the issuer is a buyer of the call option and a holder has an obligation to respond to the demand of the issuer. Thereby, a holder of a callable bond generally receives a higher yield than a usual coupon bond in exchange for taking prepayment risk. On the other hand, putable bonds give a holder the right which can sell back the bond before the maturity date. If a holder exercises the option, the issuer must buy back the bond for a put price. Then, the holder is a buyer of the put option, while the issuer is a seller. Thus, the putable bond has a lower yield than a straight bond because of the put option. In this paper, we consider the bond that the issuer and the holder are simultaneously granted the right to call and put a bond, respectively. We call this bond a game option bond.

To represent an interest rate dynamics in markets, we make use of the Generalized Ho-Lee model [1]. It is a discrete-time and an arbitrage-free term structure model of interest rate. The volatilities in each node are dependent on time and states, and hence it generalizes the Ho-Lee model [2]. For this reason, it is more flexible to model interest rate dynamics. Moreover, it is a strong point that the model has the term structure of interest rate at all nodes on the binominal lattice. We assume that the issuer and the holder decide whether to exercise the option only by taking account of the states of interest rates.

The theory of stochastic games was originated by the seminal paper of Shapley [3]. Players in stochastic games play a series of stage games that depends on a state. Using the Generalized Ho-Lee model which is a discrete-time term structure model of interest rate, we apply the stochastic game approach for pricing the game option bond as a finite time horizon problem on a binomial lattice under the risk-neutral probability. The non-arbitrage price of the game option bond is evaluated as the value of the whole game by using a dynamic programming approach [4]-[6]. We derive both the early exercise strategies and the exercise regions for players in terms of the saddle-point equilibrium strategy.

Furthermore, we report key rate duration which measures the price sensitivity of the bond to the key rate shift in a yield curve [7] [8]. Key rate durations are useful to analyze a bond with a contingent claim. We provide the intuitive numerical results for key rate durations.

The organization of this paper is as follows. In section 2, we illustrate the Generalized Ho-Lee model. In section 3, we formulate the game option bond as the stochastic game and derive the optimality equation to evaluate the non-arbitrage values of the game option bonds. In section 4, we show numerical examples for the initial values of bonds and the exercise strategies for the players, and report the key rate durations. Finally, in section 5, we conclude.

2. THE GENERALIZED HO-LEE MODEL

The Generalized Ho-Lee model is an arbitrage-free term structure model of interest rates [1]. We use recombining binomial lattice structure to describe uncertainties of interest rates. A node on the lattice is denoted by (n,i) where n is the time and i the state for $n=0,1,\cdots,N$ and $i=0,1,\cdots,n$, respectively, and N is the time horizon of the model. Let $P(n,i;T)$ represent the zero?coupon bond price at node (n,i) with remaining maturity of T periods. We call $\delta(n,i;T)$ the binomial volatility at node (n,i) with remaining maturity T, which stands for the uncertainty of interest rate on the binomial lattice. For $T=1$, we define it as

$$\delta(n,i;1)=\frac{P(n+1,i+1;1)}{P(n+1,i;1)}, 0\le n\le N-1, 0\le i\le n.$$

The Ho-Lee model, which is the first arbitrage-free term structure model of interest rates, assumes that the binomial volatilities are independent of the time and state [2] [8]. In contrast, the Generalized Ho-Lee model allows us to use the time and state dependent binomial volatilities at each node. The one-period binomial volatilities at node (n,i) are given by

$$\delta(n,i;1) = \exp\left(-2\sigma(n)\min\left(R(n,i;1),R\right)\Delta t^{\frac{3}{2}}\right) \tag{1}$$

where $\sigma(n)$ is the term structure of interest rate volatilities, $R(n,i;1)$ is the one-period yield, R is a thereshold rate, and Δt is a time interval of one period. Equation (1) signifies that the interest rate movement is lognormal when $R(n,i;1) < R$ and normal when $R(n,i;1) > R$. $\sigma(n)$ is given by the following specification:

$$\sigma(n) = (\sigma_0 - \sigma_\infty + \alpha_0 n)\exp(-\alpha_\infty n) + \alpha_1 n + \sigma_\infty \tag{2}$$

where the parameters $\sigma_0, \sigma_\infty, \alpha_0, \alpha_1, \alpha_\infty$ are estimated such that the model prices fit the swaption prices in a market. The binomial volatilities at node (n,i) with the remaining maturities T are defined by

$$\delta(n,i;T) = \delta(n,i;1)\delta(n+1,i;T-1)\left(\frac{1+\delta(n+1,i+1;T-1)}{1+\delta(n+1,i;T-1)}\right) \tag{3}$$

for $T=2,\cdots,N$. Then, the one?period zero?coupon bond prices at node (n,i) are

$$P(n,i;1)=\frac{P(0,0;n+1)}{P(0,0;n)}\prod_{k=1}^{n}\left(\frac{1+\delta(k-1,0;n-k)}{1+\delta(k-1,0;n-k+1)}\right)\prod_{j=0}^{i-1}\delta(n-1,j;1).$$

(4)

Similarly, the T-period zero-coupon bond prices are as follows:

$$P(n,i;T)=\frac{P(0,0;n+T)}{P(0,0;n)}\prod_{k=1}^{n}\left(\frac{1+\delta(k-1,0;n-k)}{1+\delta(k-1,0;n-k+T)}\right)\prod_{j=0}^{i-1}\delta(n-1,j;T).$$

(5)

We use an equal risk?neutral probability of 0.5 for the up-state and down-state. To be an arbitrage-free term structure model of interest rates, the bond prices have to satisfy the following arbitrage-free conditions:

$$P(n,i;T)=\frac{1}{2}P(n,i;1)\left[P(n+1,i;T-1)+P(n+1,i+1;T-1)\right]$$

(6)

for each $n=0,1,\cdots,N-1$, $i=0,1,\cdots,n$, and $T=1,2,\cdots,N$. When the model for bond prices is given by Equations (3)-(5), we can directly verity Equation (6) by using Equations (3)-(5).

Solving recursively Equations (1)-(5), we derive the arbitrage?free term structure model of interest rates on every node.

3. A STOCHASTIC GAME APPROACH

3.1. Callable and Putable Bonds

Callable bonds give an issuer the right to purchase back the bond for a pre-specified price (a call price) before the maturity time. If the issuer exercises the right, she pays the call price to a holder and calls back the bond. In contrast, putable bonds give a holder the right to sell back the bond at a pre-specified price (a put price) before the maturity time. If the holder exercises the right, the issuer must buy back the bond for the put price from the holder. In this paper, we consider the bond that both the issuer and the holder have the right to exercise the call and the put options, respectively. We call its bond a game option bond.

Let $t_0, t_1, \cdots, t_N$ be the time sequences on the bond, where t_0 is the initial time, $t_1, t_2, \cdots, t_{N-1}$ are the coupon times and t_N is the maturity time. We assume that each of the issuer and the holder can decide whether or

not to exercise the options at the coupon times t_n $(n \in \{n^*, n^*+1, \cdots, N-1\})$, where t_{n^*} is the first admissible exercise time. In general, we call $\{t_0, t_1, \cdots, t_{n^*}\}$ the protection period. If each of the issuer and the holder (or both) exercise at an exercisable time t_n, the holder then receives the corresponding payoffs from the issuer at the next coupon time t_{n+1}. For simplicity, we use a time index n to represent the time instead of t_n.

Let $C(n,i)$ and $P(n,i)$ denote the call and the put prices at node (n,i) for $n = n^*, n^*+1, \cdots, N-1$ and $i = 0, 1, \cdots, n$, respectively. We usually have $P(n,i) \le C(n,i)$ for every exercisable node, otherwise trivial situations arise. If both the issuer and the holder simultaneously exercise the option, we assume that the holder receives $\varphi(n,i)$ from the issuer and that $P(n,i) \le \varphi(n,i) \le C(n,i)$ for every exercisable node. In addition, if we set $P(n,i) = -\infty$, then the game option bond results in a callable bond. In contrast, if we set $C(n,i) = \infty$, then it reduces to a putable bond.

3.2. A Stochastic Game Approach

We now apply a stochastic game or a Markov game approach to the valuation of the game option bond. Let the issuer and the holder be the players of the game. Each player has two pure strategies, Exercise or Not Exercise. The holder chooses a strategy x and the issuer chooses a strategy y from the strategy set $S := \{\text{Exercise, Not Exercise}\}$ at the exercisable nodes (n,i) for $n = n^*, n^* + 1, \cdots, N-1$. The holder gains the payoff $\Psi(x, y; n, i)$ from the issuer whenever the pure strategy pair x, y is chosen at a node (n,i). If either player exercises the option at the exercisable time , for $n = n^*, n^* + 1, \cdots, N-1$, the holder receives the payoff:

$$\Psi(x, y; n, i) = \begin{cases} C(n,i) + c & \text{if the issuer exercises,} \\ P(n,i) + c & \text{if the holder exercises,} \\ \varphi(n,i) + c & \text{if both players exercise} \end{cases}$$

at the next coupon time $n+1$, where c is the fixed coupon of the bond. Although the coupon could be dependent on node (n,i), we set it as a constant for simplicity. Meanwhile, if both players don't exercise the options, the game changes from a current node (n,i) to the following nodc

$$(n+1, I_{n+1}) = \begin{cases} (n+1, i) & \text{with probability } 0.5, \\ (n+1, i+1) & \text{with probability } 0.5, \end{cases}$$

where I_{n+1} denotes a random state of interest rates at time $n+1$. Therefore, the issuer and the holder face a two-person and zero-sum stage game at every exercisable node where the payoff and the state transitions are dependent on both the state of interest rate and the strategy (action) of the two players. Then, the holder will always want to maximize her payoff, while the issuer wants to minimize the payoff, so that the rule of the strategy choice composes the exercise strategy for each player.

Given a two-person zero sum game (a matrix game) defined by a payoff matrix $\mathrm{A} \in \mathbf{R}^{m \times n} \left(m, n \in \mathbf{N} := \{1, 2, \cdots\}\right)$, where $\mathbf{R}$ stands for the set of real numbers, we define the value of the game as follows:

$$val\left[\mathrm{A}\right] := \min_{\mathbf{q} \in \Delta^n} \max_{\mathbf{p} \in \Delta^m} \mathbf{p}^{\mathsf{T}} \mathrm{A} \mathbf{q} = \max_{\mathbf{p} \in \Delta^m} \min_{\mathbf{q} \in \Delta^n} \mathbf{p}^{\mathsf{T}} \Lambda \mathbf{q} \tag{7}$$

where $\mathbf{p}$ is an m-dimensional vector representing a mixed strategies for row player, $\mathbf{q}$ an n-dimensional vector for column player and, for a positive integer $k \in \mathbf{N}$, Δ^k denote the k-dimensional unit simplex defined by

$$\Delta^k := \left\{ \mathbf{x} = \left(x_1, \cdots, x_k\right)^{\mathsf{T}} \in \mathbf{R}^k : x_i \geq 0, i = 1, \cdots, k; \sum_{i=1}^{k} x_i = 1 \right\}.$$

In Equation (7), the second equality is due to the von Neumann minimax theorem.

Let $V(n,i)$ be the value of the game option bond at node (n,i). Then, we will find the solution of the stochastic game by solving backwardly the following Optimality Equation from the maturity time $n = N$ to the initial time $n = 0$.

Optimality Equation:

$$V(n,i) = F \tag{8}$$

for $n = N$, $i = 0, 1, \cdots, n$

$$V(n,i) = \rho(n,i;1)\left\{ val\begin{bmatrix} \varphi(n,i) & P(n,i) \\ C(n,i) & E^{Q}\left[V(n+1, I_{n+1}) \middle| (n,i)\right] \end{bmatrix} + c \right\} \tag{9}$$

for $n = N-1, N-2, \cdots, n^{*}$, $i = 0, 1, \cdots, n$

$$V(n,i) = \rho(n,i;1)\left\{ E^{Q}\left[V(n+1, I_{n+1}) \middle| (n,i)\right] + c \right\} \tag{10}$$

for $n = n^{*}-1, n^{*}-2, \cdots, 0$, $i = 0, 1, \cdots, n$

where F denotes the principal of the bond, $\rho(n,i;1)$ the one-period discount factor at node (n,i) which is the one-period zero-coupon bond price based on the Generalized Ho-Lee model, Q the risk-neutral probability measure, and $E^{Q}\left[\cdot \mid (n,i)\right]$ the expectation operator conditional on node (n,i) under Q. In the exercisable node, we need to solve the two-person zero sum game defined by the following payoff matrix

$$\begin{bmatrix} \varphi(n,i) & P(n,i) \\ C(n,i) & E^{Q}\left[V(n+1, I_{n+1}) \middle| (n,i)\right] \end{bmatrix} \tag{11}$$

in order to find the exercise strategies for the players, where the holder is the row player because of a maximizer, and the issuer is the column player because of a minimizer. In the Matrix (11), for example, if the holder chooses to exercise and the issuer chooses not to exercise, then the value of the payoff matrix is $P(n,i)$. Thus, if it is possible to solve efficiently the matrix stage-game (11) for every exercisable node, we can derive simultaneously both the non-arbitrage price of the game option bond and the exercise strategies of the two players.

Next, we show the equilibrium strategies in the stage-game at each exercisable node. In general, it is well known that a two-person zero sum game has a saddle point in mixed strategies including pure strategies. However, the next result shows that a saddle point of our stochastic game exists in pure strategies.

Theorem 1. Suppose $P(n,i) < C(n,i)$ for every exercisable node. Then, the matrix stage-games played in an exercisable node always has a saddle point in pure strategies:

$$\max_{x\in S}\min_{y\in S}\Psi(x,y;n,i) = \min_{y\in S}\max_{x\in S}\Psi(x,y;n,i)$$

Furthermore, the equilibrium strategy profiles are

$$(x,y) = \begin{cases} (E,N) & \text{if } E^Q\left[V(n+1,I_{n+1})\middle|(n,i)\right] \le P(n,i) < C(n,i) \\ (N,N) & \text{if } P(n,i) < E^Q\left[V(n+1,I_{n+1})\middle|(n,i)\right] < C(n,i) \\ (N,E) & \text{if } P(n,i) < C(n,i) \le E^Q\left[V(n+1,I_{n+1})\middle|(n,i)\right] \end{cases}$$

where E denotes the pure strategy "Exercise" and N "Not Exercise".

Proof. We suppose $P(n,i) < \varphi(n,i) < C(n,i)$. If $E^Q\left[V(n+1,I_{n+1})\right] \le P(n,i) < C(n,i)$, the issuer chooses a pure strategy N because it is a dominant strategy for her (a minimizer). Then the best response strategy for the holder is a pure strategy E, and hence $(x,y)=(E,N)$. If $P(n,i)<E^{Q}\left[V(n+1,I_{n+1})\right]<C(n,i)$, a pure strategy N is a dominant strategy for both players, and $(x,y)=(N,N)$. Finally, if $P(n,i)<C(n,i)\leq E^{Q}\left[V(n+1,I_{n+1})\right]$, the holder chooses a pure strategy N which is the dominant strategy for her(a maximizer). Then, the best response strategy for the issuer is choosing a pure strategy E, and hence $(x,y)=(N,E)$.

Note that if $P(n,i)=C(n,i)$ for an exercisable node, then there is a possibility that the equilibrium strategy profile is $(x,y)=(E,E)$ because choosing to exercise is optimal for both players. However, in any case, this theorem suggests that we have only to consider a pure strategy in our stochastic game.

4. NUMERICAL EXAMPLES

This section proposes the value of the game option bond using the Generalized Ho-Lee model. We consider the bond with a 10-year maturity, the protection period of 5 years, $F=1$, $c=0.06$, and Δt = 0.25. We assume that the coupon is paid every three months and players can exercise the option at a coupon time after the protection period. For simplicity, we set $P(0,0;n)=0.05$ and $\sigma(n)=0.05$ for all n, and $C(n,i)=1.01$ and $P(n,i)=0.99$ for every exercisable node.

We first show the performance surfaces of the bond to a parallel shift in the yield curve and a change in the volatility. The yield curve is shifted in parallel from 4.5% to 5.5% every a 10 bp, and the volatility from 0% to 20% every a 2% increment. Figure 1 presents the straight coupon bond, the callable, the putable, and the game option bonds, respectively. It shows that the bond prices drop with the upward-shift in the yield curve for all bond type, because the discount factor increases. In contrast, a volatility change has different effects for each bond. The coupon bond is not influenced by a volatility change. On the other hand, as the volatility increases, the callable bond value drops, while the putable bond value rises.

Besides, we see that the game option bond drops slightly with the increase of volatilities.

We next consider the effect on the exercise regions of the issuer and the holder to the parallel shift in the parameters of interest rate. Figures 2-4 present the exercise regions of the game option bond in yield curve 4%, 5%, and 6%, respectively. In Figures 2-4, the leftmost node represents the initial time, the rightmost nodes the maturity time, the upper nodes the higher state of the interest rate, and a number in each node the game option bond value. In addition, the upper gray-areas are the holder's exercise region, while the lower enclosed areas are the issuer's one. Figures 2-4 show that the exercise region of the issuer expands with the downward shift of interest rates. Generally, if an interest rate declines after debt issues, an issuer is likely to want to refinance the loan at a lower interest rate, and hence she will exercise the call option early. In contrast, the exercise region of the holder expands with the upward shift of interest rates. This is because that the holder then can obtain higher yields by switching to other bonds, and the holder of put option will finally sell back for the issuer.

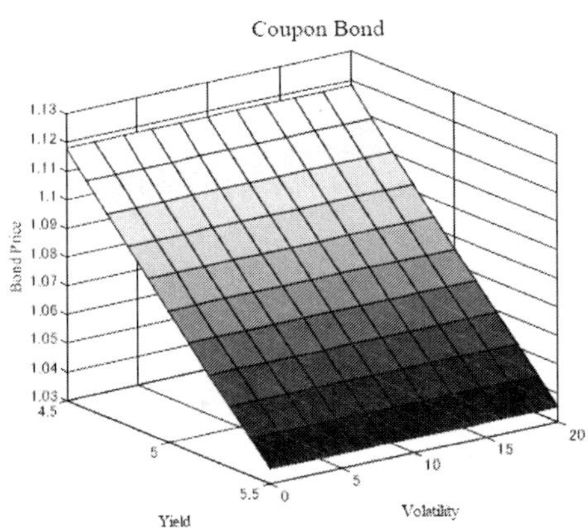

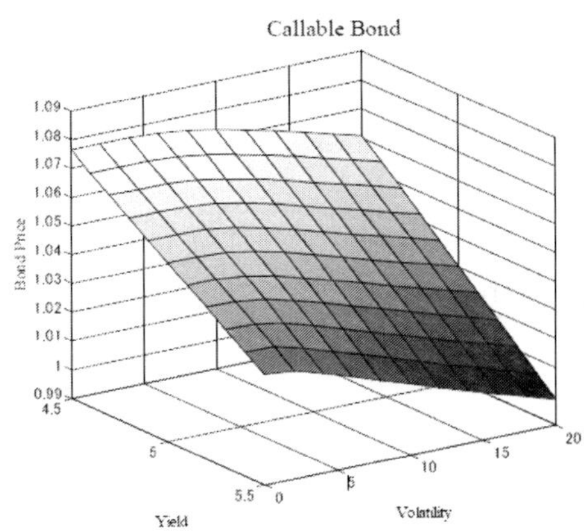

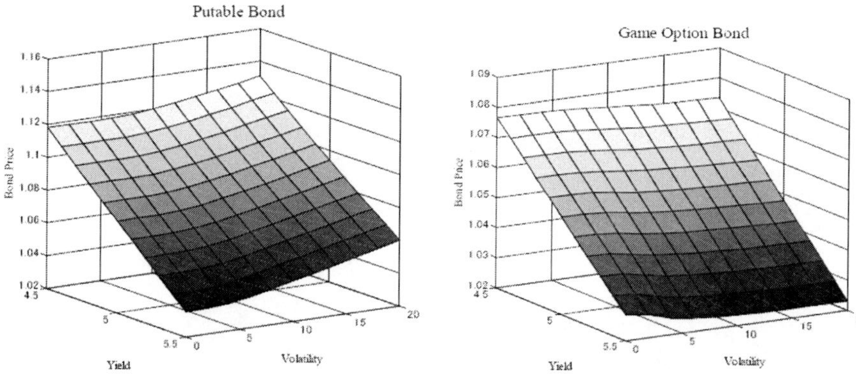

Figure 1. Surfaces of coupon bond, callable bond, putable bond, and game option bond.

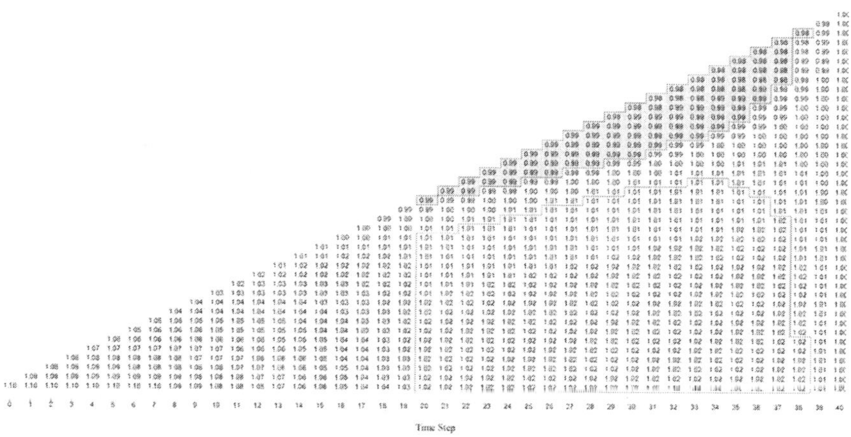

Figure 2. Exercise region of game option bond in yield curve 4%.

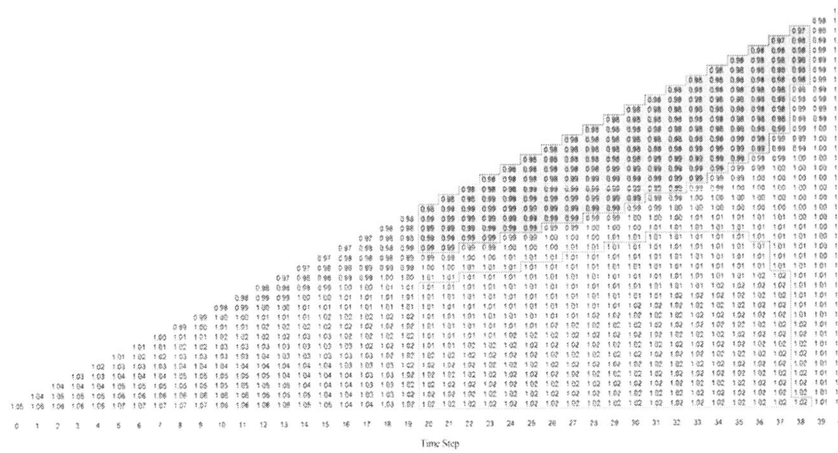

Figure 3. Exercise region of game option bond in yield curve 5%.

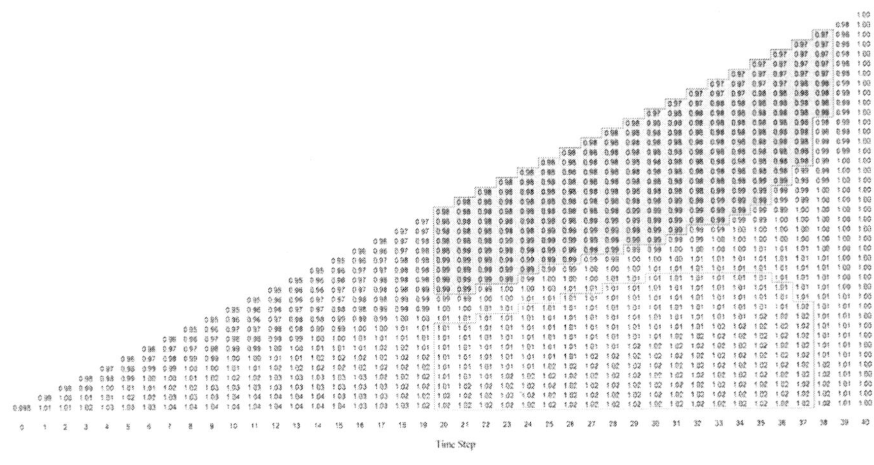

Figure 4. Exercise region of game option bond in yield curve 6%.

Figure 5 and Figure 6 represent the effect on the exercise region of the game option bond to the volatility of interest rates for 5% and 10%, respectively. From Figure 5 and Figure 6, we see that the exercise regions of the holder expand with the increase of volatility, whereas that of the issuer is not affected by the increase of volatility. We infer that the different effect is attributable to a characteristic of the Generalized Ho-Lee model, where the interest rate process is lognormal when the simulated interest rates are low and normal when the simulated interest rates are high. Thus, when the volatilities increase, the interest rate process tends to rise on the upper nodes where the interest rates are high, while the interest rate process almost never decline in the lower nodes where the interest rates are low. Finally, an increase of the interest rate volatility causes only the holder to exercise early, as shown in Figure 5 and Figure 6. In addition, it is generally known that an increase in volatilities rise a value of an option on a contingent claim. As seen from the previous Figure 1, the putable bond price rises with the increase of the volatility, and hence this is consistent with the known property of an option value.

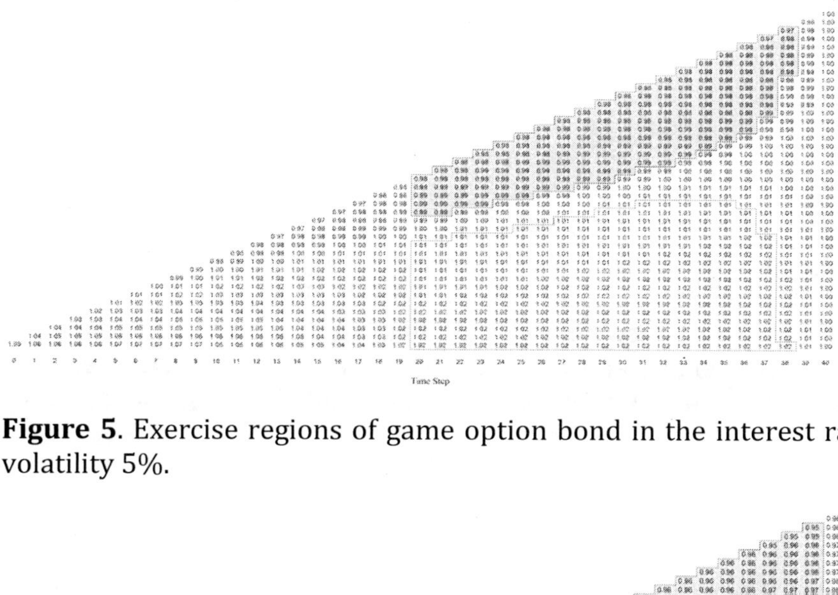

Figure 5. Exercise regions of game option bond in the interest rate volatility 5%.

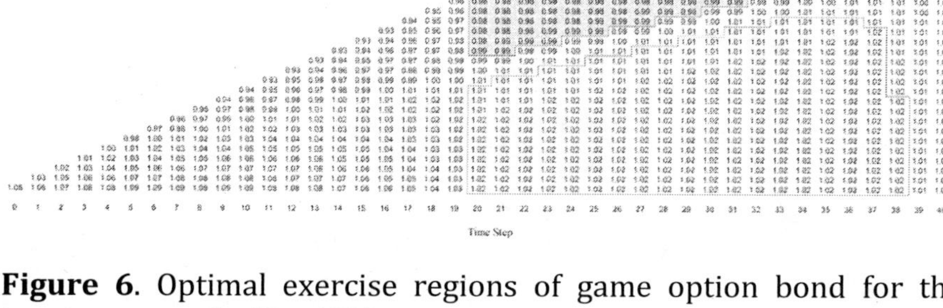

Figure 6. Optimal exercise regions of game option bond for the interest rate volatility 10%.

Key rate duration is a measure of the interest rate risk of a bond. We follow Ho [7] and report key rate duration for a variety of bonds. Key rate durations define the risk to the small change of each key rate in a yield curve. We consider a yield curve ranging from zero to ten years. To define key rate durations, we choose a set of seven key rates of three months, one, two, three, five, seven, and ten years. Key rate durations are defined as the price sensitivity to each key rate shift. The sum of key rate durations is identical to the effective duration. It can capture the risk to a yield curve change not only with parallel shift but also with steepness or curvature. We assume that the level of the key rate shift is 10 bp and the other parameters are the same as the previous given values.

Table 1 and Figure 7 compare the key rate duration profiles of the straight coupon bond, the callable, the putable, and the game option bonds. The profile shows that the callable and the game option bonds have high risk exposure to the key rate change on five years, which is the first possible exercise time. On the other hand, the coupon and the putable bonds are sensitive to the key rate shift in ten years and are alike in the key rate duration profiles. In comparing the effective durations, Table 1 shows that the option embedded in bonds have lower effective duration. It suggests that the option has an effect of reducing an impact on the interest rate risk due to the key rate change. In particular, the callable bond and the game option bond have lower price sensitivity than the straight coupon bond and the putable bond.

Table 1. Key rate durations in $C(n,i)=1.01$ and $P(n,i)=0.99$.

Bond type	Key year							*Effective duration*
	0.25	1	2	3	5	7	10	
Coupon bond	0.011	0.050	0.100	0.235	0.431	0.703	6.130	7.660
Callable bond	0.012	0.051	0.103	0.240	3.240	0.893	1.039	5.577
Putable bond	0.011	0.050	0.100	0.235	0.498	0.763	5.627	7.284
Game option bond	0.012	0.051	0.103	0.240	3.317	0.938	0.549	5.209

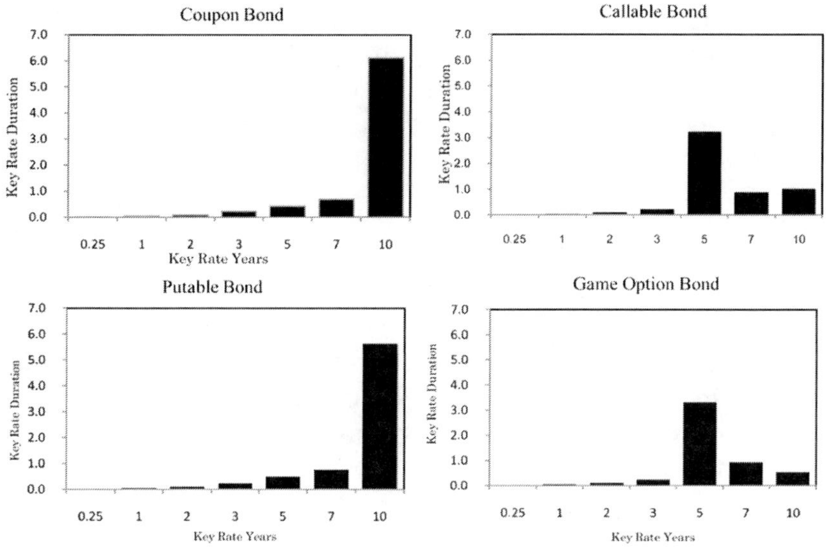

Figure 7. Key rate duration profiles in $C(n,i)=1.01$ and $P(n,i)=0.99$.

To consider the effect on game option bond value of the call option, we finally propose the key rate duration to different call price. Table 2 and Figure 8 show the key rate duration profile of the game option bond for C(n, i) = 1.01, 1.02, 1.03, and 1.04, where 1.01 is the call price forbidding the issuer to exercise at final exercisable nodes with probability 1. In general, there is a low possibility that an issuer exercises call option as the call price is higher, because the call price is considered to be the penalty for prepayment by the issuer. Table 2 and Figure 8 shows that as the call price becomes higher, the 5-year key rate duration decreases gradually, while the 10-year key rate duration increases. So the key rate duration profile of the game option bond is eventually close to that of the coupon bond with the rising call price. Accordingly, the effective duration in Table 2increases gradually.

5. CONCLUSION

In this paper, we propose a valuation of the game option bond under the Generalized Ho-Lee model. We for mulate its valuation as a stochastic game on the binomial lattice. We show that our stochastic games always possess saddle points in pure strategies at exercisable nodes. Thereby, we can efficiently solve the optimality equation to value the game option bond. Moreover, then we can obtain simultaneously both the non-arbitrage price of the game option bond and the optimal exercise strategies for the players. We provide the numerical results for the values of the bonds with a variety of contingent claims and the exercise regions for the players of the game option bond. In addition, we report the price sensitivity of the bond to a small parallel shift in some key rates by using key rate duration. The results in key rate duration show that the option embedded in a bond, particularly the call option, has the effect of reducing the exposure with respect to the interest rate change. These numerical results are useful to intuitively understand the interest rate risk in the bonds with an embedded option. However, further studies will be needed in order to value the non-arbitrage price for various interest rate derivatives related to both the credit risk and the interest rate risk [9].

Table 2. Key rate durations of game option bond for $C(n,i) = 1.01, 1.02, 1.03, \text{and } 1.04$.

Call price	Key year							Effective duration
	0.25	1	2	3	5	7	10	
1.01	0.012	0.051	0.103	0.240	3.317	0.938	0.549	5.209
1.02	0.012	0.051	0.102	0.238	3.000	0.857	1.144	5.403
1.03	0.011	0.050	0.102	0.237	2.616	0.861	1.971	5.849
1.04	0.011	0.050	0.101	0.237	2.188	0.833	2.963	6.384

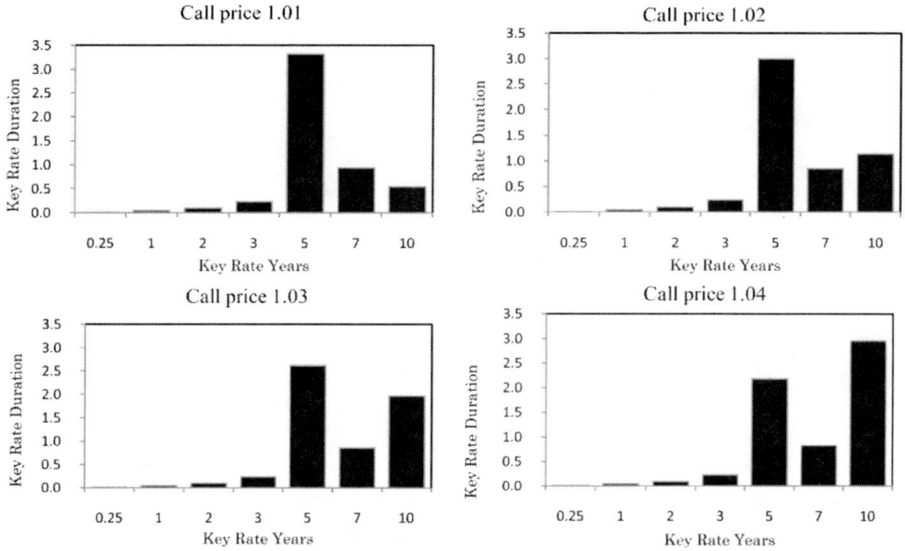

Figure 8. Key rate duration profiles of game option bond for $C(n,i) = 1.01, 1.02, 1.03, \text{ and } 1.04$.

REFERENCES

1. Ho, T.S.Y. and Lee, S.B. (2007) Generalized Ho-Lee Model: A Multi-Factor State Time Dependent Implied Volatility Function Approach. The Journal of Fixed Income, 17, 3, 18-37.

2. Ho, T.S.Y. and Lee, S.B. (1986) Term Structure Movements and Pricing Interest Rate Contingent Claims. The Journal of Finance, 41, 1011-1029.

3. Shapley, L.S. (1953) Stochastic Games. Proceedings of the National Academy of Sciences, 39, 1095-1100.

4. Ben-Ameur, H., Breton, M., Karoui, L. and L'Ecuyer, P. (2007) A Dynamic Programming Approach for Pricing Options Embedded in Bonds. Journal of Economic Dynamics & Control, 31, 2212-2233.

5. Sato, K. and Sawaki, K. (2012) The Valuation of Callable Financial Options with Regime Switches: A Discrete-Time Model. Kyoto University, RIMS Kokyuroku, 1818, Financial Modeling and Analysis, 33-46.

6. Ochiai, N. and Ohnishi, M. (2012) Pricing of the Bermudan Swaption under the Generalized Ho-Lee Model. Kyoto University, RIMS Kokyuroku, 1802, Mathematical Decision Making under Uncertainty and Ambiguity, and Related Topics, 256-262.

7. Ho, T.S.Y. (1992) Key Rate Durations Measures of Interest Rate Risks. The Journal of Fixed Income, 2, 29-44.

8. Ho, T.S.Y. and Lee, S.B. (2004) Thc Oxford Guide to Financial Modeling, Applications for Capital Markets, Corporate Finance, Risk Management, and Financial Institutions. Oxford University Press, New York.

9. Ho, T.S.Y. and Lee, S.B. (2009) A Unified Credit and Interest Rate Arbitrage-Free Contingent Claim Model. The Journal of Fixed Income, 18, 5-17.

2

CHAPTER

Parameter Dependence in Stochastic Modeling—Multivariate Distributions

Jerzy K. Filus[1], Lidia Z. Filus[2]

[1]*Department of Mathematics and Computer Science, Oakton Community College, Des Plaines, USA*

[2]*Department of Mathematics, Northeastern Illinois University, Chicago, USA*

ABSTRACT

We start with analyzing stochastic dependence in a classic bivariate normal density framework. We focus on the way the conditional density of one of the random variables depends on realizations of the other. In the bivariate normal case this dependence takes the form of a parameter (here the "expected value") of one probability density depending continuously (here linearly) on realizations of the other random variable. The point is, that such a pattern does not need to be restricted to that classical case of the bivariate normal. We show that this paradigm can be generalized and viewed in ways that allows one to extend it far beyond the bivariate or multivariate normal probability distributions class.

Keywords: Multivariate Probability Distributions, Stochastic Dependence Paradigms, Multivariate Gaussian Distributions,

Parameter Dependence Method of Construction, Conditioning, Stress, Biomedical Applications

1. INTRODUCTION

This paper can be viewed as an extension of our previous work (Filus and Filus [1]) on the bivariate Gaussian pdf structure's genesis of wide classes of newly constructed bivariate probability distributions. These distributions we constructed in our papers since 2000 up to recently (see, Filus and Filus [1] -[6] , also see Kotz, Balakrishnan and Johnson [7] pp. 217-218). Here, our explanations concerning the relation of our models to the bivariate normal were modified and we extended the topic to higher dimensional models and to the relation between our "parameter dependence method" of models construction and the multivariate normal paradigm.

It is a well-known fact that among existing multivariate probability distributions, there are no more than a few classes that are widely and successfully applied in practical stochastic modeling procedures. Typically, the underlying random variables are assumed to be independent or having an "approximately Gaussian" bivariate or multivariate distribution. The normality often is assumed even when corresponding data hardly agree with that mathematical model (showing asymmetry, for example). On the other hand, from all the multivariate distributions used in applications, the normal seems to be "the best". The reason for this is that the Gaussian models catch the stochastic relationship (mainly by a regression function) between its marginal random variables in the most natural way. We first analyze and interpret the specific way the multivariate normal density of the random vector $(X_1,\cdots,X_m)$ relates to the marginal quantities $X_1,\cdots,X_m$. Next we extend the "Gaussian pattern" to more general classes of bivariate and multivariate distributions including cases with non-Gaussian marginals. First of all we show that the common "mechanism" of the stochastic dependences both in the Gaussian distributions structure and the structure of the distributions we define, relies on the same way of conditioning. Namely, in all the considered cases, the conditional density of one random variable, say, $X_j\,(j=2,3,\cdots,m)$, given realizations, say, $x_1,\cdots,x_{j-1}$, of the marginal random variables $X_1,\cdots,X_{j-1}$ can be obtained by setting (often arbitrarily) an "initially constant" parameter, say, q_j of the density of X_j to be any continuous function $\theta_j(x_1,\cdots,x_{j-1})$ of the realizations. In this way the conditional pdf of X_j, given $x_1,\cdots,x_{j-1}$ is defined, which stands for the "source" of the stochastic dependences.

Pursuing this method successively for $j=2,3,\cdots,m$ we always arrive at a unique model. This manner, however, is characteristic for the m-variate normal where we turn an "original" normal $N(m_j, \sigma_j)$ density of X_j into the conditional density by setting a new value μ_j^* of the "affected old parameter m_j" to be the following (linear regression) function: $\mu_j^* = \mu_j + c_1x_1 + \cdots + c_{j-1}x_{j-1}$. Our main contribution is to generalize the latter function to any, not necessarily linear, function and consider not only the parameter m_j but also any other parameter of any probability density to define the corresponding conditional distributions. This is the essence of the socalled parameter dependence method. Specifically in this paper, our task will be showing more closely relation of this method to the multivariate Gaussian model construction.

In Section 2, we analyze the stochastic dependences between marginal random variables of the bivariate normal in order to point out the original version of the parameter dependence pattern next extending to other constructed bivariate probability densities. The explanation as well as the example of applications of the bivariate normal is different from that in Filus and Filus [1]. In Section 3 we present the extension of the bivariate normal pdf to the bivariate FF-normal (formerly called "pseudonormal"). In Section 4 we apply the parameter dependence method to construct the bivariate FF-Weibull (formerly "pseudo-Weibull") density in reliability framework of joint density of parallel system components life times. Comparison with other, similar, methods in the literature is presented in Section 5. In Section 6 we extend the constructions from bivariate to multivariate probability densities, first showing their relation with the multivariate normal dependences structure. Examples of the construction of multivariate FF-normal and multivariate FF-exponential ("pseudo-exponential") densities are given.

In Section 7, we point out that the "method of parameter dependence" is used in some more areas of reliability theory for different situations than we are considering. This is a part of the accelerated life testing theory where the dependence of life time distribution's parameter from a given (high) stress is investigated.

Another (fairly new) area is the "load optimization theory" sometimes associated with the load sharing phenomena analysis that we sketch in Subsection 7.2. The differences between these approaches and our theory are pointed out in 7.2.

2. THE BIVARIATE NORMAL CASE

We start with the following situation. Suppose the normally distributed random variable X_2describes an attribute of a physical or biological object, say u. Consider the (stochastic) behavior of the object u in two distinct "physical" situations. In the first situation, u is exposed to some random stress whose magnitude is described by a normally distributed random variable X_1. In the second situation we assume there is no such a stress present or the stress takes on a fixed predetermined value. The usual task here is to determine the joint distribution of X_1, X_2. Let the densities of X_1, X_2 be normal, i.e., $g_1(x_1)=N(\mu_1,\sigma_1)$, $g_2(x_2)=N(\mu_2,\sigma_2)$. [It is clear that we must assume that the value ($m_2 - k\sigma_2$) is positive for at least k = 3, in order to assure approximate positivity of the normal life-time X_2].

2.1. Simplified Biomedical Example

Imagine the following fictitious experiment whose goal is to establish the possible stochastic impact of a medication's dose change on some cancer treatment results. Suppose a person of a certain fixed age, was diagnosed with a kind of cancer. Assume that one of the significant characteristics of that kind of cancer is a tumor with a size X_2. During a given time period T after the patient was diagnosed, a specific medication was administered. Also suppose that this medication was routinely administered in the past, and that the average dose is estimated (or fixed) to be m_1 milligrams per kilo of weight daily. Assume that, originally, the known (either measured or estimated) average size of the tumor is, say, m_2 millimeters and after the period T of treatment the tumor size X_2 is measured again and its negative or positive increment $X_2 - \mu_2$ is statistcally confronted with the dose X_1 of the medication administered.

We assume that the goal of the underlying experiment is to make a prediction on effect $X_2 - \mu_2$ of the treatment when the dose is changed from its "original level" $X_1 = \mu_1$ to a level $X_1 \neq \mu_1$. Randomness of the doze X_1 may be justified when only "historical data" are analyzed and then extrapolated for a larger population of cases not yet recorded. In the case of extrapolation of historical data for a larger population we assume that the only information one possess on the applied in the past dose X_1 is its probability distribution, which is the Gaussian $N(\mu_1, \sigma_1)$ with given values for both the parameters m_1, σ_1. Also, for $X_1 = \mu_1$ the tumor size X_2 (after the treatment) is assumed to be a random variable having a normal $N(\mu_2, \sigma_2)$density, where $X_2 - \mu_2$ is the value of the tumor "increment" under the treatment characterized by the dose level m_1. For any other applied doze $X_1 = x_1$ the, associated with a single patient, value $(x_1 - m_1)$ statistically affects the change in the tumor size $X_2 - m_2$ i.e., the treatment result. The word "statistically" here means that the impact of a nonzero quantity $(x_1 - m_1)$ ("the dose is not the standard one") on the (former) probability density $N(m_2, \sigma_2)$ of the tumor size X_2 realizes through affecting the value of the mean m_2 rather than directly affecting the numerical value x_2 of X_2.

If we were interested in finding the joint probability distribution of X_1, X_2 it is enough to determine the conditional density $g_2(x_2|x_1)$ of $X_2|X_1$, since the marginal density of X_1 is not changing.

In accordance with the "linear regression rule", the dependence of the (new) expected value μ_2^* (so new probability density $N(\mu_2^*, \sigma_2^*)$) of the tumor size X_2 on the event $X_1 = x_1$, is determined by the familiar functional relationship:

$$E\left(X_2 | X_1 = x_1\right) = \mu_2^* = \mu_2 + a\left(x_1 - \mu_1\right) \tag{1}$$

where a = $r(\sigma_2/\sigma_1)$ and r is the (linear) correlation coefficient of the variables X_1, X_2.

This approach directly leads to the determination of the conditional density of the random variable X_2 given any realization $X_1 = x_1$. It is a well-known fact that the conditional density $g_2(x_2|x_1)$ is, again, normal and

$$g_2\left(x_2|x_1\right) = \frac{1}{\sigma_2\sqrt{2\pi\left(1-\rho^2\right)}} \exp\frac{-\left(x_2 - \mu_2 - a\left(x_1 - \mu_1\right)\right)^2}{2\sigma_2^2\left(1-\rho^2\right)} \quad ,(2)$$

i.e., the $N\left(\mu_2 + a\left(x_1 - \mu_1\right), \sigma_2\sqrt{1-\rho^2}\right)$ density in x_2.

The joint density $g(x_1,x_2)$ of the random variables X_1, X_2 is given by the usual arithmetic product $g_2(x_2|x_1)g_1(x_1)$.

2.2. Remark

In the example above, one can reinterpret the "response random variable" X_2 to be for example the patient's "residual life-time", or blood pressure, or level of some important chemical in the blood (such as cholesterol). In such cases the mathematics of the problem would remain the same.

Note the obvious fact that the tumor size X_2 does not have a physical influence on the medication dose X_1 so that the original marginal pdf $g_1(x_1)$ remains the same. However, the stochastic dependence between X_1, X_2 is mutual, since, in general, $g_1(x_1|x_2) \neq g_1(x_1)$.

It is well known that the actual problem with the bivariate normal density construction is to get to the conditional density (2), which fully represents the underlying stochastic dependence of random variable X_2 on X_1.

Our claim is that the above paradigm for the stochastic dependence (characteristic for the bivariate Gaussians) can be extended to other classes of bivariate and multivariate distributions (see Filus and Filus [3] [4]).

3. THE FF-NORMAL (PSEUDONORMAL) EXTENSION

Historically, people relied on the nice symmetry in the stochastic dependence of X_1 and X_2 when using their joint bivariate normal distribution. This kind of symmetry (i.e., both marginal and both conditional distributions are normal and both sides regression functions are linear) can only be achieved with the linear regression functions as described above (1). However, are linear regression functions really the only functions that one can successfully apply within this framework? Assuming that the function $m_2(X_1)$ is any continuous function in X_1, one obtains a wide and interesting extension of the class of bivariate normal densities. We called this class FF-normal (previously named "pseudonormal", see [2] and also [7]). In this case the parameter σ_2 can as well become a continuous function of the stress X_1 (X_1 may have a "stress" interpretation in a very wide sense). This stress may change the parameter σ_2 of the (normal) density of, say, the "life-time" X_2 into another value $\sigma_2^* = \sigma_2^*(x_1)$ that depends on the particular realization x_1 of the random variable X_1. The price for such a wide generalization is loss of the, mentioned above, symmetry (the marginal of X_2 ceases to be normal) but the gains are considerable. Anyway, the bivariate normal remains to be a special cases of the FF-normals.

With the bivariate FF-normal densities of (X_1, X_2) we can use general continuous $m_2(x_1)$ and $\sigma_2(x_1)$ functions, and, performing similar calculations as above, we find, rather surprisingly, that $g_2(x_2|x_1)$ is once more a regular normal density in x_2.

Consider now the following situation with a bivariate FF-normal distribution in which the "physical" interpretation of the underlying random variables can now be more general than above. Let u_1, u_2 be two objects (or phenomena) which are characterized by the random variables X_1, X_2 respectively. If the objects are physically separated then the random variables X_1, X_2 are assumed to be independent, having normal pdfs $g_1(x_1)=N(\mu_1,\sigma_1)$ and $g_2(x_2)=N(\mu_2,\sigma_2)$ respectively. When the objects physically interact (or rather only u_1 physically impacts u_2), then the corresponding joint FF-normal density $g(x_1, x_2)$ of the random vector (X_1, X_2) is given by the usual product formula:

$g(x_1,x_2)=g_1(x_1)g_2(x_2|x_1)$ with the invariant marginal density $g_1(x_1)=N(\mu_1,\sigma_1)$.

For the conditional density of $X_2|x_1$ we have: $g_2(x_2|x_1)=N(\mu_2^*(x_1),\sigma_2^*(x_1))$.

The functions $\mu_2^*(x_1),\sigma_2^*(x_1)$ are "formed" from the "no-stress" parameters m_2, σ_2 of X_2's density.

More explicitly, one obtains the bivariate FF-normal pdf in the form:

$$g(x_1,x_2)=\frac{1}{\sqrt{2\pi}\sigma_1}\exp\frac{-(x_1-\mu_1)^2}{2\sigma_1^2}\frac{1}{\sqrt{2\pi}\sigma_2^*(x_1)}\exp\frac{-(x_2-\mu_2^*(x_1))^2}{2\sigma_2^{*2}(x_1)}$$

where $\mu_2(x_1)=E[X_2|x_1]$ is the (in general, nonlinear) "regression function", and $[\sigma_2^*(x_1)]^2=\mathrm{Var}[X_2|x_1]$ is the conditional variance obtained from the "previous" variance σ_2^2.

In particular, one may consider the "nonlinear regression function"

$$E[X_2|x_1]=\mu_2(x_1)=\mu_2+a^*(x_1-\mu_1)+A(x_1-\mu_1)^n$$

with arbitrary real parameters a^* and A, $n=2,3,\cdots$.

Realize that in the case A = 0 and $\sigma_2(x_1) = \sigma_2\sqrt{1-\rho^2}$ we obtain the regular bivariate Gaussian density. The coefficient A of the term $A(x_1 - \mu_1)^n$ may be considered as a nonlinear "correction" of the regular Gaussian (linear) regression. The main purpose of that correction is to enhance the accuracy in various modeling procedures. For some type of asymmetric data, especially interesting may be the "non-symmetric" "quadratic" case n = 2.

Equally important is the "symmetric" "cubic" case when n = 3. Also interesting is the case where a^*= 0 and A ≠ 0. Here, the linear correlation coefficient is zero, while the (nonlinear) dependence may be quite "heavy". Another interesting case (which may be combined with those above) is when we choose the standard deviation function as belonging to the "parametric class" of "quadratic functions" defined by $\sigma_2^*(x_1) = k\sigma_2\left(1 + r^2 x_1^2\right)$ with k positive real. It is rather obvious that the idea of the construction of bivariate normal (and FF-normal as well) can be extended to other probability distributions such as exponential, gamma, Weibull, lognormal, etc.

Generally speaking, the essence of the construction method is that for any pair of ("initially independent") random variables X_1 and X_2 with given probability densities $g_1(x_1, \omega)$ and $g_2(x_2;\Theta)$ respectively, one can simply "declare" some parameter (or a vector parameter), say Θ, of the density $g_2(x_2;\Theta)$ to be dependent on the values of the other random variable X_1. This means that when $X_1 = x_1$ we may assume that $g_2(x_2, \Theta(x_1)) = g_2(x_2|x_1)$ is the "affected by x_1" distribution of the random variable X_2 (given the event $X_1 = x_1$ occurred, with probability density $g_1(x_1)$).

Then the joint density of the pair (X_1, X_2) is always $g(x_1, x_2) = g_2(x_2, \Theta(x_1))g_1(x_1)$.

This situation is especially natural if we consider X_2 to be the life-time of an object and X_1 is the stress put on it.

Roughly, one can say that the construction method of bivariate distributions, presented above is an extension of the method used in the construction of the bivariate normal.

4. RELIABILITY EXAMPLE

Consider a 2-component (say u_1, u_2) parallel system reliability setting in which X_1, X_2 represent the components' life-times (see Barlow and Proschan [8]). We start with the situation where the system's components act separately. We call that pattern the "laboratory conditions". In this latter case the components are physically separated and consequently their life-times (represented by the, statistically estimated, "baseline probability densities" $g_1(x_1)$ and $g_2(x_2)$ respectively), are stochastically independent. When the two components are installed into the system, they start to interact. Assume that during that interaction some irregularities in the work of component u_1 cause corresponding changes in u_2's inner physical structure. This increases the hazard rate of that physically affected component u_2. Such physical phenomena are then "responsible" for the occurrence of stochastic dependence in the "in-system" component life-times X_1 and X_2.

One can also imagine this situation as follows. During the two components' "in-system" performance, component u_1 creates a situation in which component u_2 is "constantly bombarded" by a string of harmful "micro-shocks" (see Filus and Filus [5]). Each such micro-shock causes a corresponding "micro-damage" in the affected component u_2's physical constitution. We also assume that these micro-damages in component u_2's inner physical structure "cause" some corresponding "micro-changes in the original (baseline) failure rate" (and, in parallel, in the corresponding probability distribution) of its life-time X_2. After a, possibly long, time period X_1 of such interaction all these micro-damages cumulate their effects. As a result of this accumulation, the overall change in the corresponding "hazard rate function" will become significant. To describe formally the change in the hazard rate function we have chosen to consider corresponding changes in its parameter(s). In what follows we present a particular bivariate model for a 2-component system reliability which we called FF-Weibullian (formerly "pseudo-Weibullian" in Filus and Filus [4]).

Suppose the lifetimes of the components u_1 and u_2 in "laboratory conditions" are independent and distributed according to the Weibull density random variables X_1 and X_2.

Let $g_k(x_k) = \lambda_k \alpha_k x_k^{\alpha_k - 1} \exp - \lambda_k x_k^{\alpha_k}$ be the pdf of X_k (k = 1, 2).

Here, for k = 1, 2, we have the "vector parameter" $\theta_k = (\lambda_k, a_k)$.

Next consider the components u_1 and u_2 as acting within the system.

Let the resulting (changed) values λ_2^*, σ_2^* of the parameters of the (original) pdf $g_2(x_2; l_2, a_2)$ be determined by the following continuous functions of x_1:

$\lambda_2^* = \lambda_2^*(x_1)$ and $\alpha_2^* = \alpha_2^*(x_1)$.

One then obtains the wide class of bivariate FF-Weibullian densities:

$$g(x_1, x_2) = g(x_1) g_2(x_2 | x_1) = \lambda_1 \alpha_1 x_1^{\alpha_1 - 1} \exp\left(-\lambda_1 x_1^{\alpha_1}\right) \lambda_2^*(x_1) \alpha_2^*(x_1) x_2^{\alpha_2^*(x_1) - 1} \exp\left(-\lambda_2^*(x_1) x_2^{\alpha_2^*(x_1)}\right) \tag{3}$$

where, for ease of computation, we recommend to apply as "sub-model" the following family of "parameter functions":

$\lambda_2^*(x_1) = \lambda_2 \left(1 + A x_1^r\right)$ and $\alpha_2^*(x_1) = s\alpha_2$ with parameters A, r, s positive reals.

In particular, s may depend on x_1.

Another analytically interesting "sub-model" is given by:

$\lambda_2^*(x_1) = \lambda_2 \exp\left[A x_1^r\right]$ with A and r real, and $\alpha_2^*(x_1) = s\alpha_2 (1 + c x_1)$ with s > 0, c 3 0.

Note that both factors $g_1(x_1)$ and $g_2(x_2|x_1)$ of the joint density $g(x_1, x_2)$ given by (3) are Weibullian densities. In particular, $g_2(x_2|x_1)$ is Weibullian with respect to the argument x_2 alone.

For the simpler FF-exponential example, see [1].

5. NOTICE ON SIMILAR INVESTIGATIONS IN THE LITERATURE

A parallel and basically independent path of investigation, which also has its roots in the bivariate normal distribution's dependence paradigm, is present in the literature under the key word "conditioning".

This method, used in the construction of numerous multivariate probability distributions, was extensively developed mostly since around 1987. See, for example, Arnold, Castillo and Sarabia [9] with citations. Also consider Castillo and Galambos [10].

The underlying method (by numerous authors called the "conditioning method") relies on imposing conditional structure X|Y and Y|X on, given in advance, "baseline" probability densities f(x; A) and g(y; B) of some ("initially independent") random variables X and Y respectively, where A and B are scalar or vector parameters. The two conditional densities are defined as we did above, i.e.

$$g(y|x) = g(y; B(x)) \text{ and } f(x|y) = f(x; A(y))$$

where A(y) and B(x) are continuous functions of realizations of the random variables Y, X respectively.

In this case the task is to find two proper (unknown) marginal densities for the bivariate probability distribution of (X, Y) which are, as a rule, not unique and sometimes do not exist.

Despite similarities this method essentially differs from ours. In our case, instead of the two conditional densities g(y|x) and f(x|y), we define only one, say, g(y|x), but together with the marginal f(x).

Pursuing this way we always directly obtain a unique model simply as the product of the two (known) densities.

In such a way, we have obtained a wide class of bivariate densities which is essentially disjoint from the class obtained by that alternative method. Also, the physical interpretation of the, so defined, conditional densities differs in the two approaches.

However, both approaches are devoted to the same purpose which is to extend of the paradigm of the bivariate normal in stochastic modeling. Nevertheless, using the conditioning method it is very difficult to construct the multivariate distributions of any higher than two dimensions.

Practically that method reduces to the bivariate cases while the method we present has a remarkable easiness of construction of probability distributions of, actually, arbitrary finite dimension. There is, namely, a recurrence procedure which allows to construct any j-th dimensional pdf based on corresponding (j – 1)-th dimensional pdf $(j=3,4,\cdots)$ constructed "at an earlier stage". That procedure was also used in Filus and Filus [11] for the construction of discrete time stochastic processes.

The next section is devoted to the construction of multivariate distributions for any arbitrary finite dimension.

6. METHOD OF PARAMETER DEPENDENCE FOR MULTIVARIATE PROBABILITY DISTRIBUTIONS CONSTRUCTION

For the construction, mentioned in the title, we successively use the simple recurrence method that yields the j-th dimensional probability density, given the (j – 1)-th one. Realize, that (for j = 3) we have already defined the 2- dimensional densities $g^2(x_1, x_2)$ by means of the products $g_1(x_1)g_2(x_2|x_1)$, where each underlying conditional density was given by $g_2(x_2|x_1) = g_2(x_2, q_2(x_1))$.

So the "first step" is already done. Suppose now that we have at our disposal the (j – 1)-th dimensional (j ³ 3) pdf, say $g^{j-1}(x_1,x_2,\cdots,x_{j-1})$ of the random vector $(X_1,X_2,\cdots,X_{j-1})$. The task of obtaining the joint density $g^j(x_1,x_2,\cdots,x_j)$ of the random vector $(X_1,X_2,\cdots,X_{j-1},X_j)$ always reduces to defining the conditional density $g_j(x_j|x_1,x_2,\cdots,x_{j-1})$, given any univariate baseline pdf $h_j(x_j,\theta_j)$ by the method of parameter dependence. Assuming that originally q_j is a constant parameter, we define the conditional pdf $g_j(x_j|x_1,x_2,\cdots,x_{j-1})$ by setting (according to the new physical situation of "being in the system" together with the other j – 1 objects):

$$g_j(x_j|x_1,x_2,\cdots,x_{j-1})=h_j(x_j,\theta_j^*(x_1,x_2,\cdots,x_{j-1}))$$

Now, the "new" value θ_j^* of the parameter is a continuous function of the realizations ("multi-stresses") $(x_1,x_2,\cdots,x_{j-1})$ of the random vector $(X_1,X_2,\cdots,X_{j-1})$.

The j-dimensional pdf of the random vector $(X_1,X_2,\cdots,X_{j-1},X_j)$ one obtains simply as the product:

$$g^j(x_1,x_2,\cdots,x_j)=g_j(x_j|x_1,x_2,\cdots,x_{j-1})g^{j-1}(x_1,x_2,\cdots,x_{j-1})$$

The latter pdf becomes the basis for identical construction of the (j + 1)-dimensional pdf and so on.

We then stop the procedure once j + 1 = m, where m is the total dimension of the considered (maximal) random vector, say, $(X_1,\cdots,X_m)$.

Since the analogy with the construction of each j-dimensional normal pdf $(j=3,4,\cdots)$ is not as straightforward as in the bivariate case, we found it beneficial to show this analogy closer, by reducing the normal's construction to the "diagonal case". Let us start with recalling that any normally distributed random vector, say, $X=(X_1,\cdots,X_j)$, $j=2,3,\cdots,m$ is obtainable from the random vector $Z=(Z_1,\cdots,Z_j)$ by an affine transformation

$$X^{\mathrm{T}} = AZ^{\mathrm{T}} + \mu^{\mathrm{T}}, \tag{4}$$

where the random variables $Z_1,\cdots,Z_j$ are independent and each having the standard normal N(0, 1) pdf. A is an arbitrary j × j matrix with real entries (here, without losing generality, we restrict ourselves to nonsingular matrices A, only) and $\mu=(\mu_1,\cdots,\mu_j)$ is an arbitrary fixed vector in R^j. The symbol ^{T} denotes the usual matrix transpose. Recall that every matrix A can be decomposed as

$$A = MB, \tag{5}$$

where B is a lower triangular and M is an orthogonal matrix. From (5) we obtain that any nonsingular lower triangular matrix B can be represented as the product:

$$B = M^{\mathrm{T}}A, \tag{6}$$

where A is an arbitrary nonsingular matrix. If we replace representation (4) of the random normal vector X by the following representation

$$Y^{\mathrm{T}} = BZ^{\mathrm{T}} + \mu^{\mathrm{T}}, \tag{7}$$

then we replace the arbitrary random vector X by an arbitrary "triangular" random vector Y related to X by:

$$(Y-\mu)^{\mathrm{T}} = M^{\mathrm{T}}(X-\mu)^{\mathrm{T}}, \tag{8}$$

where M^T is an arbitrary orthogonal transformation.

Since the two zero-expectation random vectors X – μ, Y – μ are obtained one from another by an isometry (here, rotation) M^T in the Euclidean space R^j, they may be considered as representing the same "stochastic data" expressed in two different (but still rectangular) coordinate systems. So from a stochastic viewpoint the "difference" between the random vectors X and Y is inessential and we can consider the random vector Y as an "arbitrary normal" ("with accuracy to the rotation" M^T).

Collecting all the above, we will consider the normal random vector Y, given by (7), where matrix B is any lower triangular matrix and Z is the standard normal j-vector. Write (7) in the form:

$$\begin{aligned}
Y_1 &= b_{11}Z_1 + \mu_1 \\
Y_2 &= b_{21}Z_1 + b_{22}Z_2 + \mu_2 \\
&\vdots \\
Y_{j-1} &= b_{j-1,1}Z_1 + \cdots + b_{j-1,j-1}Z_{j-1} + \mu_{j-1} \\
Y_j &= b_{j,1}Z_1 + \cdots + b_{j,j-1}Z_{j-1} + b_{j,j}Z_j + \mu_j \quad j \to m
\end{aligned} \tag{7*}$$

where m ^{3} j is the actual dimension of the constructed (final) random vector, say $(X_1,\cdots,X_m)$ (if m = □, one defines, in effect, a stochastic process with time j).

Considering the first –1 lines in (7*) as a system of linear equations, one obtains all $Z_1,\cdots,Z_{j-1}$ as linear combinations of $Y_1,\cdots,Y_{j-1}$. Substituting these solutions back into (7*) one obtains the following form:

$$\begin{aligned}
Y_1 &= c_{11}Z_1 + \mu_1 \\
Y_2 &= c_{22}Z_2 + \left(c_{21}Y_1 + \mu_2\right) \\
Y_3 &= c_{33}Z_3 + \left(c_{31}Y_1 + c_{32}Y_2 + \mu_3\right) \\
&\vdots \\
Y_{j-1} &= c_{j-1,j-1}Z_{j-1} + \left(c_{j-1,1}Y_1 + \cdots + c_{j-1,j-2}Y_{j-2} + \mu_{j-1}\right)
\end{aligned} \tag{9}$$

Realize that transformation (9) is easily reversible.

Assuming that realizations $y_1,\cdots,y_{k-1}$ of the random variables $Y_1,\cdots,Y_{k-1}$ are known, we obtain for each $k=1,\cdots,j$:

$$Z_k = \frac{Y_k - \left(c_{k,1}y_1 + \cdots + c_{k,k-1}y_{k-1} + \mu_k\right)}{c_{kk}} \tag{10}$$

where from the above assumed nonsingularity we have $c_{kk} \neq 0$. From (10) it follows that the conditional density of each Y_k, given the values $y_1,\cdots,y_{k-1}$, is normal and for the corresponding conditional expectation we have

$$E\left[Y_k \middle| y_1,\cdots,y_{k-1}\right] = \mu_k + c_{k,1}y_1 + \cdots + c_{k,k-1}y_{k-1}$$

while for the (constant) conditional variance we obtain

$$\mathbf{Var}\left[Y_k \middle| y_1,\cdots,y_{k-1}\right] = c_{kk}^2$$

To adopt the above procedure to our concept of "baseline" T_j versus "in system" Y_j random variables, replace in (9) the independent standard random variables $Z_1, Z_2, \cdots$ by independent random variables, say, $T_1, T_2, \cdots$, where each T_k has the ("baseline") normal $N(\mu_k, s_k)$ pdf $(k = 1, 2, \cdots)$.

Replace transformation (7) by

$$Y = BT^{\mathrm{T}}, \tag{11}$$

where $T = (T_1, \cdots, T_j)$. Using this change, (10) will be replaced by the following inverse transformation:

$$T_k = \frac{Y_k - (c_{k,1} y_1 + \cdots + c_{k,k-1} y_{k-1})}{c_{kk}} \tag{10*}$$

This yields the conditional pdf of $Y_k | y_1, \cdots, y_{k-1}$ to be the normal

$$N\left(\mu_k + c_{k,1} y_1 + \cdots + c_{k,k-1} y_{k-1}, |c_{kk}| \sigma_k\right)$$

Finally, the general pattern of "creation" of any successive j-variate normal pdf can be explained as follows.

Given are the first j – 1 lines of transformation (11) in the form:

$$
\begin{aligned}
Y_1 &= c_{11}T_1 \\
Y_2 &= c_{22}T_2 + c_{21}Y_1 \\
Y_3 &= c_{33}T_3 + \left(c_{31}Y_1 + c_{32}Y_2\right) \\
&\vdots \\
Y_{j-1} &= c_{j-1,j-1}T_{j-1} + \left(c_{j-1,1}Y_1 + \cdots + c_{j-1,j-2}Y_{j-2}\right)
\end{aligned}
\tag{12}
$$

for some $j-1=1,2,\cdots$. (Realize that the joint normal pdf $g^{j-1}\left(y_1,y_2,\cdots,y_{j-1}\right)$ of the random vector $\left(Y_1,Y_2,\cdots,Y_{j-1}\right)$ was defined in j – 1 "previous" steps. In particular for j – 1 = 1, it is the univariate normal $N(\mu_1, |c_{11}|s_1)$ density of the variable Y_1.)

We may assume that the next baseline random variable T_j, originally having the $N(\mu_j, s_j)$ pdf, is incorporated to the "system" by transforming

$$T_j \to Y_j \middle| Y_1,\cdots,Y_{j-1}$$

This transformation is thought of as adding to (12) the following j-th line:

$$Y_j = c_{j,j}T_j + \left(c_{j,1}Y_1 + \cdots + c_{j,j-1}Y_{j-1}\right) \tag{13}$$

["Physically" this could mean that the variables $Y_1,\cdots,Y_{j-1}$ "become" explanatory ("stresses") for the "new" variable Y_j obtained from T_j that (originally) was independent from these stresses].

From (13) one can determine the conditional pdf of Y_j, given any realization $\left(y_1,\cdots,y_{j-1}\right)$ of the random vector $\left(Y_1,\cdots,Y_{j-1}\right)$, as the following normal pdf in y_j:

$$N\left(\mu_j + c_{j,1}y_1 + \cdots + c_{j,j-1}y_{j-1}, \left|c_{jj}\right|\sigma_j\right).$$

Thus, as the j-th "object" (originally independent from the "system" and characterized by the random quantity T_j) was "put into the system" the quantity T_j turns to the quantity Y_j and, in parallel, the parameters μ_j and s_j of its normal density are turned into $\mu_j + c_{j,1}y_1 + \cdots + c_{j,j-1}y_{j-1}$ and $\sigma_j\left|c_{jj}\right|$, respectively, while normality is preserved.

Clearly, the new value μ_j^* of the (conditional) expectation became the continuous (here linear) function

$$\mu_j^*\left(y_1, \cdots, y_{j-1}\right) = \mu_j + c_{j,1}y_1 + \cdots + c_{j,j-1}y_{j-1}$$

of realizations $y_1, y_2, \cdots, y_{j-1}$ while unfortunately the new value of the standard deviation does not depend on $y_1, y_2, \cdots, y_{j-1}$ but remains constant even if multiplied by the specific, determined by the "system", number c_{jj}. This can be "made up" if we allow the number c_{jj} in (13) to be dependent on $y_1, y_2, \cdots, y_{j-1}$, but then the so obtained multivariate FF-normal distribution ceases to be normal since (13) ceases to be linear.

As we have shown also in multivariate cases, the origin of the "parameter dependence method for the construction", lies in the construction of the multivariate normal distributions. Recall that having defined the conditional pdf $g_j\left(y_j \middle| y_1, \cdots, y_{j-1}\right)$ and the joint pdf $g^{j-1}\left(y_1, \cdots, y_{j-1}\right)$ we automatically have the joint pdf $g^j\left(y_1, \cdots, y_j\right)$ as the simple arithmetic product of the two. In the case just considered, all the densities $g^j(y_1,\cdots,y_j), j=1,2,\cdots,$ are (arbitrary with the accuracy to the rotations in R^j) multivariate normal.

Preserving the general spirit of the multivariate normal pdf derivation, let us extend all the Equations (13) for $j=2,3,\cdots,m$ by allowing the translations $c_{j,1}Y_1+\cdots+c_{j,j-1}Y_{j-1}$ to be any nonlinear continuous function of $Y_1,\cdots,Y_{j-1}$ and replacing the constant $c_{j,j}$ by any continuous function of the same variables. Now, for any $j=2,3,\cdots,m$ (13) may be rewritten into the following "triangular" (see Filus, Filus and Arnold [12]) form:

$$Y_j = \Phi_j\left(Y_1,\cdots,Y_{j-1}\right)T_j + \Psi_j\left(Y_1,\cdots,Y_{j-1}\right) \tag{13*}$$

where $\Phi_j()$ and $\Psi_j()$ are arbitrary continuous functions and. $P(\Phi_j(Y_1,\cdots,Y_{j-1})\neq 0)=1$. From (13*) we obtain its inverse:

$$T_j = \frac{Y_j - \Psi_j\left(Y_1,\cdots,Y_{j-1}\right)}{\Phi_j\left(Y_1,\cdots,Y_{j-1}\right)} \tag{13**}$$

and then for each observation $(y_1,\cdots,y_{j-1})$ of $(Y_1,\cdots,Y_{j-1})$ the conditional pdf of $Y_j|y_1,\cdots,y_{j-1}$ as follows:

$$g_j\left(y_j|y_1,\ \ldots\ ,y_{j-1}\right)=N\left(\mu_j+\Psi_j\left(y_1,\cdots,y_{j-1}\right);\sigma_j\left|\Phi_j\left(y_1,\cdots,y_{j-1}\right)\right|\right) \tag{14}$$

It is clear that the sequence of the densities (14) $(j=2,3,\cdots,m)$ together with the normal initial density $g_1(y_1)$ of Y_1 uniquely determines the m-variate FF-normal pdf of the random vector.

Remarkably, this non-normal density has its natural representation as the product of m normal densities:

$$g(y_1,\cdots,y_m)=g_1(y_1)\prod_{j=2}^{m}g_j\left(y_j\middle|y_1,\cdots,y_{j-1}\right) \qquad (15)$$

However, the marginal pdfs of $Y_2,\cdots,Y_m$ are not normal anymore.

The main conclusion which follows the considerations in Sections 6.2 - 6.4 may be stated as: There is a generic relationship that associates the construction method of the parameter dependence with the stochastic dependence structure present within the multivariate normal distribution of any dimension.

As an example of this relationship realize that the transformations (13) and (13*), when applied to the independent normal random variables $T_j\ (j=1,2,\cdots,m)$, define multivariate normal and FF-normal pdfs respectively. They produce other m-variate probability distributions if the normality assumption for T_j is dropped.

Let now $T_j\ (j=1,2,\cdots,m)$ be independent random variables all having the standard exponential pdf:

$$f_j(t_j)=\exp\left[-t_j\right].$$

Applying to the random vector $(T_1,\cdots,T_m)$ transformation (13*) and then (13**) for $j=1,2,\cdots,m$, one obtains the joint density $g(y_1,\cdots,y_m)$ of the resulting random vector $(Y_1,\cdots,Y_m)$ to be given by the product of m factors (15), where according to first row of (12) we have $g_1(y_1)=(1/|c_{11}|)\exp\left[-y_1/|c_{11}|\right]$ and according to (13**)

$$g_j\left(y_j\middle|y_1,\cdots,y_{j-1}\right)=\frac{1}{\left|\Phi_j(y_1,\cdots,y_{j-1})\right|}\exp\frac{-\left(y_j-\Psi_j(y_1,\cdots,y_{j-1})\right)}{\left|\Phi_j(y_1,\cdots,y_{j-1})\right|}$$

where the latter is the two parameter exponential density with respect to y_j for $j = 2,3,\cdots,m$.

Another interesting case of the m-variate FF-Weibullian pdf can be obtained by applying transformations (13*) to m independent Weibullian random variables. An even more general class of FF-Weibullians one obtains using the pseudopower transformations (see Filus and Filus [4]) instead of the pseudoaffine (13*) which actually is a special case of the pseudopower.

All these distributions (including the m-variate normal) can as well be obtained by direct use of the "parameter dependence pattern" which produces more m-variate models than the considered above transformations. On the other hand existence of the defining transformations facilitates an underlying statistical analysis and simulations.

7. OTHER PARAMETER DEPENDENCE PARADIGMS IN THE LITERATURE

Some paradigms, applied in the reliability literature, are exactly those of the "parameter dependence" that we describe in this paper. However, in most of the cases they are not directly related to the problem of construction of multivariate probability distributions (so, also are different from the "conditioning" procedures in [9] [10]; see above, Section 5). There are two such subjects that we discuss in the following.

7.1. The Accelerated Life Testing

When testing the life times of some high reliability products, the stresses usually encountered such as temperature, humidity, voltage sometimes are kept on significantly higher than usual levels in order to make the life times shorter than they are in normal conditions. The so obtained data (a "sample") is then extrapolated into those (hypothetical) life times that would, possibly, be obtained under the regular values of the stresses. Existence of rules, that associate the products' life times with values of the stresses applied, is necessary for performing proper extrapolations. Several such rules, typically known as the Arrhenius or Eyring (see, Meeker and Escobar [13] , Nelson [14] and [15] an internet source) relationships, are based on physical and chemical considerations on the rates of some chemical reactions that give rise to a given unit's failure. The obtained models, in general, allow determining the ratio of life times of the same product under higher and under normal temperature or, in the case of the Eyring model, some other physical quantities that play the role of the stresses (see, [13], formula (18.5) page 476). Methods like that (i.e., the so called SAFT models [13]) directly relate the (life) times by means of a simple coefficient called the "acceleration factor".

Unfortunately, with this method the simplicity often comes along with inaccuracy of the predictions. Other methods apply the "Proportional Hazards Relationships" known also as Cox Model (see Cox [16]) which instead of times relate hazard rates.

More recently ([15]), the relationships between the life times under different stresses are related indirectly through their probability distributions via distribution's parameter in that way that considers distribution's parameter as a function of a given stress (mostly temperature, humidity, pressure, voltage). Those relationships, even if considered in a different context, obey the same paradigm (of the parameter dependence) as that considered in this paper in association with the construction of the multivariate probability distributions.

In what follows we discuss the differences.

1) The generality of the "parameter dependence theory" we built in this paper, is significantly higher than the very special case applied to the accelerated life testing theory. There are three reasons for that.

Firstly, in our approach the subject of constructing conditional probability distributions is not limited to the life testing, and not even to the "stress-life time" pattern only. The range of applications of our theory is very wide, including many biomedical (see, Collett [17]) and econometric relationships, (see, Filus and Filus [11] also Filus, Filus and Krysiak [18]).

Secondly, in the paradigm we consider, the relation between a parameter and stress (or any other random quantity) is given by an arbitrary continuous function, while the number of such functions applied in association with the accelerated life time testing is very limited. Actually, the functions are restricted to few "models" such as the Arrhenius, Eyrie, inverse power law, log-linear, and not many more (see, for example, the Eyring-Weibull model in [15], Section 5). Those models were obtained from physical and chemical considerations that are only valid for some simple failure (degradation) mechanisms while very often real failure mechanism is too complicated to be analyzed that way.

Our idea is to omit the complicated physical or chemical phenomena that often are poorly understood and to apply two steps purely empirical approach.

Speaking roughly, the first step is an "educated guess" (for choice of a proper function) and the second is statistical verification of this guess.

Thirdly, in our theory we may consider an arbitrary parameter of an arbitrary probability distribution as a stress dependent, while, according to our knowledge (see, for example, [15]), the only life time distributions so far considered in the accelerated life testing are exponential, Weibullian, and lognormal (normal), and for each the distribution only one parameter is taken under consideration.

2) Besides the generality (of the constructed conditional distributions) our concept also differs with regard to the purpose. Namely, independently of the conditional distributions construction, we also have the construction of bivariate and multivariate probability distributions such as the FF-normal, FF-exponential, FF-Weibullian, FF-gamma and other (for comparison with similar "conditioning methods" of construction present in the literature, see Section 5).

The construction of high dimension multivariate distributions based on parameter dependence can easily be extended to Markovian and non-Markovian (still simple!) stochastic processes (see Filus and Filus [11]). The latter constructions seem to be rather unique in the literature.

7.2. Load Control and Load Sharing

1) Other than the accelerated life testing subject, where the "parameter dependence paradigm" is applied, is a set of problems centered around the notion of "load optimization" (see Filus [19] [20] , Levitin and Amari [21], Nourelfath and Yalaoui [22] and others). This relatively new topic can be described as follows. Some working systems, such as cargo transporting trucks, trains, electric power lines, highways, computer processors, or other systems supporting varying amounts of load, require control of that load. On the one hand more load yields more gain, but, on the other hand, load (as stress) increment yields a corresponding increment of the system's failure rate which, in turn, depends on certain parameters [20]. The optimization problem can be formulated in several ways ([20] [21]) but, in all cases, the main idea is to balance between an expected gain that follows a good (load) transportation and the loss in reliability of the transporting medium which decreases the overall gain. Thus, for some (optimal) value of the load to be found, the pure expected gain earned by the system is maximal. In this framework the relationship between stress (the load) and some parameters of the system life time probability distribution (failure rate) is vital for finding a proper model. As examples of such relations, the power and exponential functions of the load were chosen in [20].

2) Similar application of the parameter dependence pattern also occurs when "load sharing phenomena" takes place. Suppose that we have a parallel system supporting a load such as several engines aircraft or two electric power lines. Failure of any system's component may cause the total load to be redistributed among fewer components, so that the load on each of them increases by some predictable value. Now we may encounter either the load optimization problem [21] or simply the task of determination of system's life time probability distribution (see Freund [23] for the exponential case as well as Lu [24] and Filus [25] for Weibullian and lognormal cases). In all these cases the parameter dependence pattern is involved or it is desirable to apply it to get a deeper insight into the underlying stochastic phenomena.

Remark. As a final remark, let me mention the relationship between the parameter dependence presented in this paper, and the stochastic dependence based on models initiated in 1961 by Freund [23]. Besides some similarity the most basic difference lies in the fact that in the Freund scheme the system components act independently until the first failure. In general, the components successive failures cause the total load to be shared by fewer and fewer remaining components, affecting their failure rates (via the parameters).

Quite opposite to that, in the models we introduce, the component interactions take place only when the components work. Any failure of a system component stops its influence on the remaining components' life times. Therefore, the two paradigms, the "Freund's load sharing" and our "parameter dependence", are "disjoint" and in a sense "complementary". In reality, both (physical) phenomena may take place at the same time and it seems to be quite possible in the future to construct stochastic models (i.e., multivariate probability distributions) that would obey both paradigms.

Nevertheless, we stress the generic relation of all the multivariate probability distributions based on the parameter dependence with the multivariate Gaussians.

REFERENCES

1. Filus, J.K. and Filus, L.Z. (2013) A Method for Multivariate Probability Distributions Construction via Parameter Dependence. Communications in Statistics: Theory and Methods, 42, 716-721.

2. Filus, J.K. and Filus, L.Z. (2000) A Class of Generalized Multivariate Normal Densities. Pakistan Journal of Statistics, 16, 11-32.

3. Filus, J.K. and Filus, L.Z. (2007) On New Multivariate Probability Distributions and Stochastic Processes with Systems Reliability and Maintenance Applications. Methodology and Computing in Applied Probability, 9, 426-446,

4. Filus, J.K. and Filus, L.Z. (2006) On Some New Classes of Multivariate Probability Distributions. Pakistan Journal of Statistics, 1, 21-42.

5. Filus, J.K. and Filus, L.Z. (2008) On Multicomponent System Reliability with Microshocks-Microdamages Type of Components' Interaction. Proceedings of the International Multiconference of Engineers and Computer Scientists, Lecture Notes in Engineering and Computer Science, Hong Kong, 19-21 March 2008, 1945-1951.

6. Filus, J.K. and Filus, L.Z. (2010) Weak Stochastic Dependence in Biomedical Applications. American Institute of Physics Conference Proceedings 1281, Numerical Analysis and Applied Mathematics, III, 1873-1876.

7. Kotz, S., Balakrishnan, N. and Johnson, N.L. (2000) Continuous Multivariate Distributions (Vol. 1). 2nd Edition, J. Wiley & Sons, Inc, New York, 217-218.

8. Barlow, R.E. and Proschan, F. (1975) Statistical Theory of Reliability and Life Testing. Holt, Rinehart and Winston, New York.

9. Arnold, B.C., Castillo, E. and Sarabia, J.M. (1999) Conditional Specification of Statistical Models. Springer Series in Statistics, Springer Verlag, New York.

10. Castillo, E. and Galambos, J. (1990) Bivariate Distributions with Weibull Conditionals. Analysis Mathematica, 16, 3-9.

11. Filus, J.K., Filus, L.Z. (2008) Construction of New Continuous Stochastic Processes. Pakistan Journal of Statistics, 24, 227-251.

12. Filus, J.K., Filus, L.Z. and Arnold, B.C. (2010) Families of Multivariate Distributions Involving "Triangular" Transformations. Communications in Statistics—Theory and Methods, 39, 107-116.

13. Meeker, W.Q. and Escobar, L.A. (1998) Statistical Methods for Reliability Data. John Wiley & Sons, Inc., New York.

14. Nelson, W. (1990) Accelerated Testing: Statistical Models, Test Plans and Data Analysis. Wiley, New York.

15. (2013)

16. Cox, D.R. (1972) Regression Models and Life Tables (with Discussion). Journal of the Royal Statistical Society, B74, 187-220.

17. Collett, D. (2003) Modeling Survival Data in Medical Research. 2nd Edition, Chapman @ Hall/CRS A CRC Press Company, London, New York, Washington, D.C.

18. Filus, J.K., Filus, L.Z. and Krysiak, Z. (2013) Analytical Statistical and Simulation Models Utilized in Modeling the Risk in Finance. 5th International Conference on Risk Analysis, Tomar, Portugal, 30th May-1st June.

19. Filus, J.K. (1986) A Problem in Reliability Optimization. Journal of the Operational Research Society, 37, 407-412.

20. Filus, J.K. (1987) The Load Optimization of a Repairable System with Gamma Distributed Time-to-Failure. Reliability Engineering, 18, 275-284.

21. Levitin, G. and Amari, S.V. (2009) Optimal Load Distribution in Series-Parallel Systems. Reliability Engineering and System Safety, 94, 254-260.

22. Nourelfath, M. and Yalaoui, F. (2012) Integrated Load Distribution and Production Planning in Series-Parallel Multistate Systems. Reliability Engineering and System Safety, 106, 138-145.

23. Freund, J.E. (1961) A Bivariate Extension of the Exponential Distribution. Journal of the American Statistical Association, 56, 971-977.

24. Lu, J. (1989) Weibull Extensions of the Freund and Marshal-Olkins Bivariate Exponential Models. IEEE Transactions on Reliability, 38, 615-619.

25. Filus, J.K. (1991) On a Type of Dependencies between Weibull Life times of System Components. Reliability Engineering and System Safety, 31s, 267-280.

3

CHAPTER

Optimal Portfolio Strategy with Discounted Stochastic Cash Inflows

Charles I. Nkeki

Department of Mathematics, Faculty of Physical Sciences, University of Benin, Benin City, Nigeria

ABSTRACT

This paper examines optimal portfolios with discounted stochastic cash inflows (SCI). The cash inflows are invested into a market that is characterized by inflation-linked bond, a stock and a cash account. It was assumed that inflationlinked bond, stock and the cash inflows are stochastic and follow a standard geometric Brownian motion. The variational form of Merton portfolio strategy was obtained by assuming that the investor chooses constant relative risk averse (CRRA) utility function. The inter-temporal hedging terms that offset any shock to the SCI were obtained. A closed form solution to our resulting non-linear partial differential equation was obtained.

Keywords: Optimal Portfolio; Stochastic Cash Inflows; Inflation-Linked Bond; Variational Form; Intertemporal Hedging Terms

1. INTRODUCTION

This paper consider the optimal portfolio strategies with valuing expected discounted cash inflows. It was assumed that the underlying assets and cash inflow process follow a standard geometric Brownian motion. The investment of the investor's SCI into a cash account, an inflationlinked bond and a stock was considered. The SCI and stock price were correlated with inflation and stock market risk.

In a related literature, [1] studied an optimal investment problem in a continuous-time framework where the interest rate follows the Cox-Ingersoll-Ross dynamics. They obtained a closed form solution for the optimal investment strategy under a complete market framework. They assumed that the investor chooses CRRA utility function. [2] considered the process of finding the optimal portfolio, optimal consumption and the efficient frontier for a small agent in an economy. They also considered financial market that composed of two sources of uncertainties: an m-dimensional Brownian motion and a continuous time Markov chain. They found theoretically, that the regimes of an economy have a significant impacts on the portfolio and consumption decision as well as on the efficient frontier of a small investor. [3] studied the present value and overcome the difficulty of independence by reversing the order of the cash flow. He found that similar recursive formulas for the present value is also applicable to the future value of the expected returns. [4] considered the problem of constructing a portfolio of finitely many assets whose returns are described by a discrete joint distribution. They proposed a new portfolio optimization model involving stochastic dominance constraints on the portfolio return. [5] studied how portfolio composition changes with individual wealth. [6] considered the present value of expected future contribution into the pension fund. [7] investigated the stochastic dynamics of depository financial institution assets, liabilities and capital under the influence of macroeconomic factors. They adopted dynamic programming techniques in their optimization process. They provided an analysis of the economic aspects of the depository financial institution modeling. [8] studied a portfolio problem of a fund manager who wants to maximize the expected utility of his terminal wealth in a complete financial market. He found that the optimal portfolio is formed by three components: a speculative, an hedging, and a preference-free hedging component. He obtained a close form solution to the asset allocation problem. [9] considered the optimal portfolio selection problem with portfolio constraints. They derived the general utility function using the martingale approach. They found that using CRRA utility function, optimal policies can be obtained explicitly when there are minimum capital requirements. ([10,11]) studied the optimal portfolio management in the accumulation phase of a defined contribution pension scheme. They obtained portfolio values with hedging strategies for a pension plan member. In this paper, we consider the optimal portfolio and investment strategies involving cash inflow valuation over time. We obtain explicitly, analytical solution to our resulting HJB equation. The risk-free rate is assume to be deterministic.

The remaining parts of the paper is structured as follows: In Section 2, we described the structure of the financial market; In Section 3, we consider the dynamics of the discounted present value of expected SCI process; In Section 4, we consider the wealth process of the investor; In Section 5, we consider the valuation of the discounted present value of expected SCI process of the investor as well as the sensitivity of the present value of the SCI; In Section 6, we consider the optimization program and optimal portfolio and optimal solution of the investor's wealth using CRRA utility function; Section 7 concludes the paper.

2. THE MODEL

Let (Ω, F, P) be a probability space. Let

$\mathrm{F}(F)=\{F_t : t\in[0,T]\}$, where

$F_t=\sigma(S(s),I(s):s\leq t)$. The Brownian motions

$W(t)=(W^I(t),W^S(t))$ is a 2-dimensional process, defined on a given filtered probability space

$(\Omega, F, \mathrm{F}(F), \mathrm{P})$, $t \in [0,T]$, where P is the real world probability measure, t the time period, T the terminal time period, $W^I(t)$ is the Brownian motion with respect to source of uncertainty arising from inflation and $W^S(t)$ is the Brownian motion with respect to source of uncertainty arising from the stock market. It is assumed that the market is arbitrage-free, complete and continuously open between time period 0 and T.

Financial Model

The dynamics for the cash account with the price $Q(t)$ at time t is given by

$$\begin{aligned} &\mathrm{d}Q(t) = rQ(t)\,\mathrm{d}t, \\ &Q(0) = 1; \end{aligned} \tag{1}$$

where r is the short term interest rate. The stock price $S(t)$ at time t is given by the dynamics:

$$\begin{aligned} &\mathrm{d}S(t) = S(t)\left(\mu \mathrm{d}t + \sigma_1 \cdot \mathrm{d}W(t)\right), \\ &S(0) = s_0; \end{aligned} \tag{2}$$

where, μ is the expected growth rate of stock price,

$\sigma_1 = \left(\rho\sigma_S, \sigma_S\sqrt{1-\rho^2}\right)$ and $0 < \rho < 1$. σ_S is the volatility of stock. The price of the inflation-linked bond $B(t, I(t))$ is given by the dynamics:

$$\begin{aligned} &\mathrm{d}B(t, I(t)) = B(t, I(t))\left[\left(r + \sigma_I\theta_I\right)\mathrm{d}t + \sigma_B \cdot \mathrm{d}W(t)\right], \\ &B(0, I(0)) = b; \end{aligned} \tag{3}$$

where, $\sigma_B = (\sigma_I, 0)$ is the volatility of inflation-linked bond, θ_I is the market price of inflation risk, $I(t)$ is the inflation index at time t and has tha dynamics:

$$dI(t) = qI(t)dt + \sigma_I I(t)dW^I(t),$$

where q is the expected rate of inflation, which is the difference between nominal interest rate, r real interest rate $\overline{r}$ and σ_I is the volatility of inflation index. Since the market is complete, we have that

$$\Sigma := \begin{pmatrix} \sigma_B \\ \sigma_1 \end{pmatrix} = \begin{pmatrix} \sigma_I & 0 \\ \rho\sigma_S & \sigma_S\sqrt{1-\rho^2} \end{pmatrix}, \tag{4}$$

$$\lambda := (\sigma_I \theta_I, \mu - r)'. \tag{5}$$

Therefore, the market price of market risk is given by

$$\theta := \Sigma^{-1}\lambda = \begin{pmatrix} \theta_I \\ \theta_S \end{pmatrix} = \begin{pmatrix} \theta_I \\ \dfrac{\mu - r - \rho\sigma_S\theta_I}{\sigma_S\sqrt{1-\rho^2}} \end{pmatrix}, \tag{6}$$

where, θ_S is the market price of stock market risk. The exponential process

$Z(t):=\exp\left(-\theta'\cdot W(t)-\frac{1}{2}\|\theta\|^2 t\right), 0\le t\le T,$ is assumed to be a martingale. We now define the state-price density function by

$$\Lambda(t):=\frac{Z(t)}{Q(t)}=Z(t)\exp(-rt),$$
$$0\le t\le T.$$

3. DYNAMICS OF STOCHASTIC CASH INFLOWS

The dynamics of the stochastic cash inflows with price process, $D(t)$ is given by

$$dD(t)=D(t)\left(kdt+\sigma_D'\cdot dW(t)\right),$$
$$D(0)=D_0; \tag{7}$$

where, $\sigma_D=\left(\sigma_1^D,\sigma_2^D\right)'$ is the volatility of the cash inflows and k is the expected growth rate of the cash inflows. σ_1^D is the volatility arising from inflation and σ_2^D is the volatility arising from the stock market. Figure 1 presents the simulated diffusion paths of (7).

Figure 1 was obtained by setting $k=0.099$,

$\sigma_1^D=0.25$, $\sigma_2^D=0.36$, $D_0=100$, $dt=\frac{t-t_0}{nt}$, $t=10$, $t_0=0$ and $nt=1000$.

Applying It$\hat{o}$ lemma to (7), we obtain

$$D(t)=D_0\exp\left(\left(k-\frac{1}{2}\|\sigma_D\|^2\right)t+\sigma_D'\cdot W(t)\right). \tag{8}$$

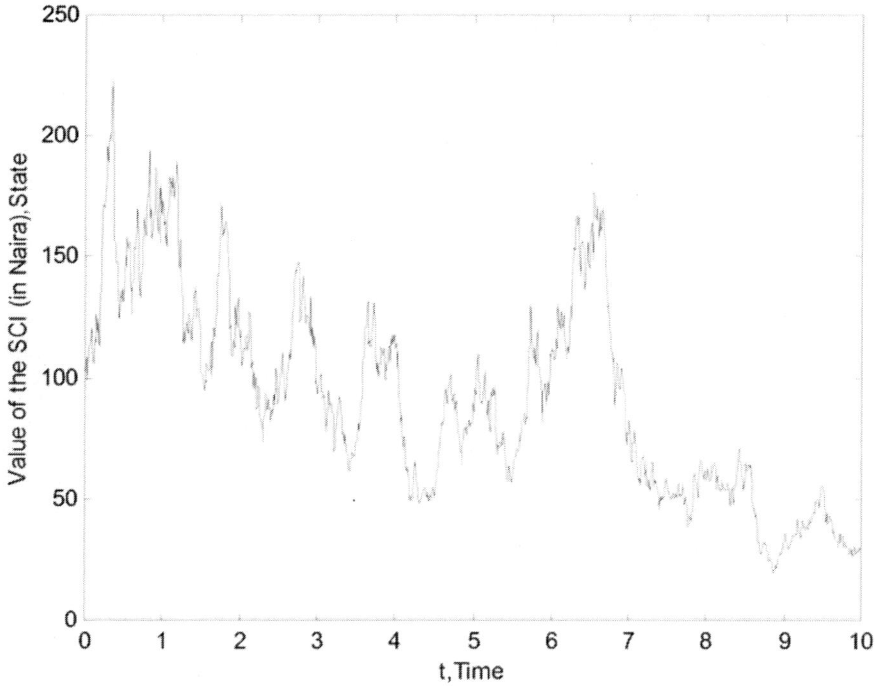

Figure 1. Simulated diffusion paths of the stochastic cash inflows.

4. THE WEALTH PROCESS

Let $X(t)$ be the wealth process and $\Delta(t)=(\Delta_I(t),\Delta_S(t))'$ be the portfolio value at time t where $\Delta_I(t)$ is the portfolio value in inflation-linked bond and $\Delta_S(t)$ is the portfolio value in stock at time t. Then, $\Delta_0(t)=1-\Delta_I(t)-\Delta_S(t)$ is the portfolio value in cash account at time t. Therefore, the dynamics of the wealth process is given by

$$
\begin{aligned}
dX(t) &= X(t)\Delta_S(t)\frac{dS(t)}{S(t)} \\
&+ X(t)\Delta_I(t)\frac{dB(t,I(t))}{B(t,I(t))} \\
&+ X(t)(1-\Delta_I(t)-\Delta_S(t))\frac{dQ(t)}{Q(t)} + D(t)dt, \\
X(0) &= x.
\end{aligned}
\tag{9}
$$

Substituting (1), (2) and (3) into (9), we obtain

$$
\begin{aligned}
dX(t) &= \left(X(t)(r + \lambda \cdot \Delta(t)) + D(t)\right)dt \\
&\quad + X(t)(\Sigma\Delta(t))' \cdot dW(t), \\
X(0) &= x.
\end{aligned}
\tag{10}
$$

5. THE DISCOUNTED VALUE OF SCI

In this section, we determine the value of expected discounted SCI.

Definition 1:

The discounted value of the expected future SCI is defined as

$$
\Psi(t) = E_t\left(\int_t^T \frac{\Lambda(u)}{\Lambda(t)} D(t)\mathrm{d}u\right)
\tag{11}
$$

where, $E_t = E(\cdot \mid F_t)$ is the conditional expectation with respect to the Brownian filtration $\{F_t\}_{t>0}$ and

$\Lambda(t) = Z(t)\exp(-rt)$ is the stochastic discount factor which adjusts for nominal interest rate and market price of risks for stock and inflation-linked bond.

Proposition 1:

Suppose $\Psi(t)$ is the discounted value of the expected future SCI, then

$$\Psi(t)=\frac{D(t)\left(\exp\left((k-r-\sigma_D\cdot\theta)(T-t)\right)-1\right)}{k-r-\sigma_D\cdot\theta}. \tag{12}$$

Proof. By definition 1, we have that

$$\Psi(t)=D(t)E_t\left(\int_t^T\frac{\Lambda(u)D(u)}{\Lambda(t)D(t)}\mathrm{d}u\right) \tag{13}$$

Applying change of variable on (13), we have

$$\Psi(t)=D(t)E\left(\int_0^{T-t}\frac{\Lambda(\tau)D(\tau)}{\Lambda(0)D(0)}\mathrm{d}\tau\right) \tag{14}$$

Applying parallelogram law and martingale principles on (14), we have

$$E\left(\frac{\Lambda(\tau)D(\tau)}{\Lambda(0)D(0)}\right)=\exp\left((k-r-\sigma_D\cdot\theta)\tau\right).$$

Therefore,

$$\Psi(t)=D(t)\left(\int_0^{T-t}\exp\left((k-r-\sigma_D\cdot\theta)\tau\right)\mathrm{d}\tau\right). \tag{15}$$

Integrating, we have

$$\Psi(t)=\frac{D(t)\left(\exp\left(\left(k-r-\sigma_D\cdot\theta\right)\left(T-t\right)\right)-1\right)}{k-r-\sigma_D\cdot\theta}. \tag{16}$$

Figure 2 was obtained by setting $D_0=100$, k = 0.099, r = 0.04, $\sigma_1^D=0.25$, $\sigma_2^D=0.36$, $\mu=0.09$, $\sigma_S=0.4$, $\sigma_I=0.3$, $\rho=0.6$ and $\theta_I=0.08$.

At $t=0$, we obtain the present discounted value of future SCI to be

$$\Psi(0)=\Psi_0=\frac{D_0\left(\exp\left(\left(k-r-\sigma_D\cdot\theta\right)T\right)-1\right)}{k-r-\sigma_D\cdot\theta}. \tag{17}$$

If $r+\sigma_D\cdot\theta>k$ and we allow $T\rightarrow\infty$ i.e.,

$$\lim_{T\to\infty}\Psi_0=\lim_{T\to\infty}\left[\frac{D_0\left(\exp\left(\left(k-r-\sigma_D\cdot\theta\right)T\right)-1\right)}{k-r-\sigma_D\cdot\theta}\right],$$

$$r+\sigma_D\cdot\theta>k$$

$$=\frac{D_0}{r+\sigma_D\cdot\theta-k}, r+\sigma_D\cdot\theta>k.$$

For a deterministic case, we have the present discounted value future deterministic cash inflows to be

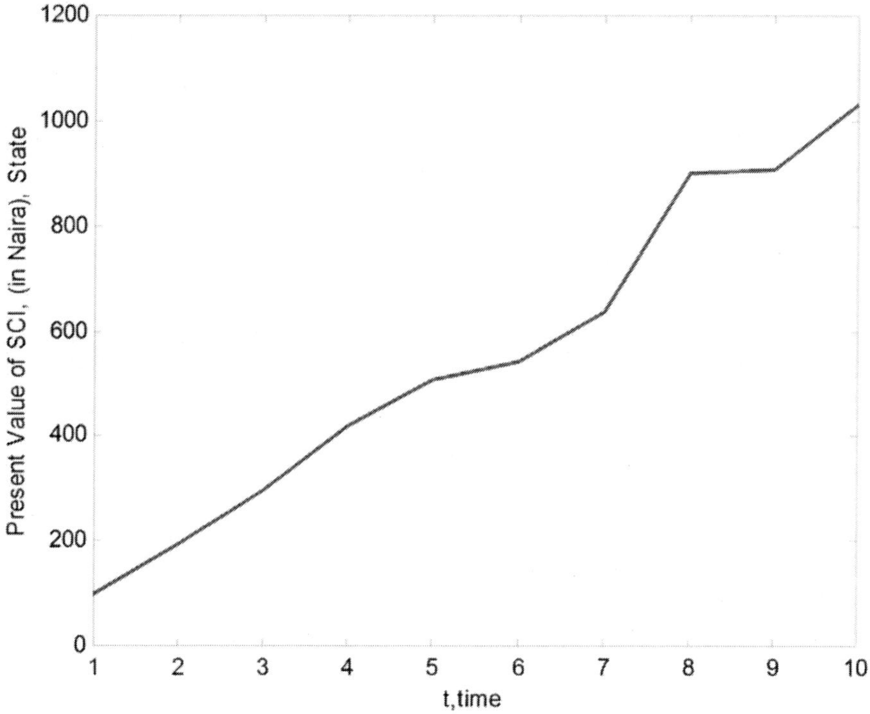

Figure 2. The flow of the discounted value of SCI.

$$\Psi_0 = \frac{D_0\left(\exp\left((k-r)T\right)-1\right)}{k-r}. \tag{18}$$

Hence,

$$\lim_{T\to\infty} \Psi_0 = \lim_{T\to\infty}\left[\frac{D_0\left(\exp\left((k-r)T\right)-1\right)}{k-r}\right], r > k$$

$$= \frac{D_0}{r-k}, r > k.$$

This shows that as $T \to \infty$, Ψ_0 converges to $\frac{D_0}{r+\sigma_D \cdot \theta - k}$, provided $r+\sigma_D \cdot \theta > k$.

For the deterministic case, it shows that as $T \to \infty$, Ψ_0 converges to $\frac{D_0}{r-k}$, provided $r > k$.

Figure 3 was obtained by setting $D_0 = 100$, k = 0.099, r = 0.01, $\sigma_1^D = 0.25$, $\sigma_2^D = 0.36$, $\mu = 0.09$, $\sigma_S = 0.4$, $\sigma_I = 0.3$, $\rho = 0.6$ and $\theta_I = 0.08$.

Figure 2 represents the flow of the discounted value of stochastic cash inflow in the investment at time t and **Figure 3** represents the present value of discounted future SCI at time t.

We now consider the sensitivity analysis of Ψ_0. Proposition 2 establishes this fact.

Proposition 2:

Let $\beta = k - r - \sigma_D \cdot \theta$, then

$$\frac{\partial \Psi_0}{\partial T} = D_0 \exp(\beta T).$$

Proof. The results follow by taking the partial derivatives of Ψ_0 with respect to T, D_0, r, σ_D and k, respectively.

$$\frac{\partial \Psi_0}{\partial D_0} = \frac{1}{\beta}\left(\frac{1}{D_0}\frac{\partial \Psi_0}{\partial T} - 1\right);$$

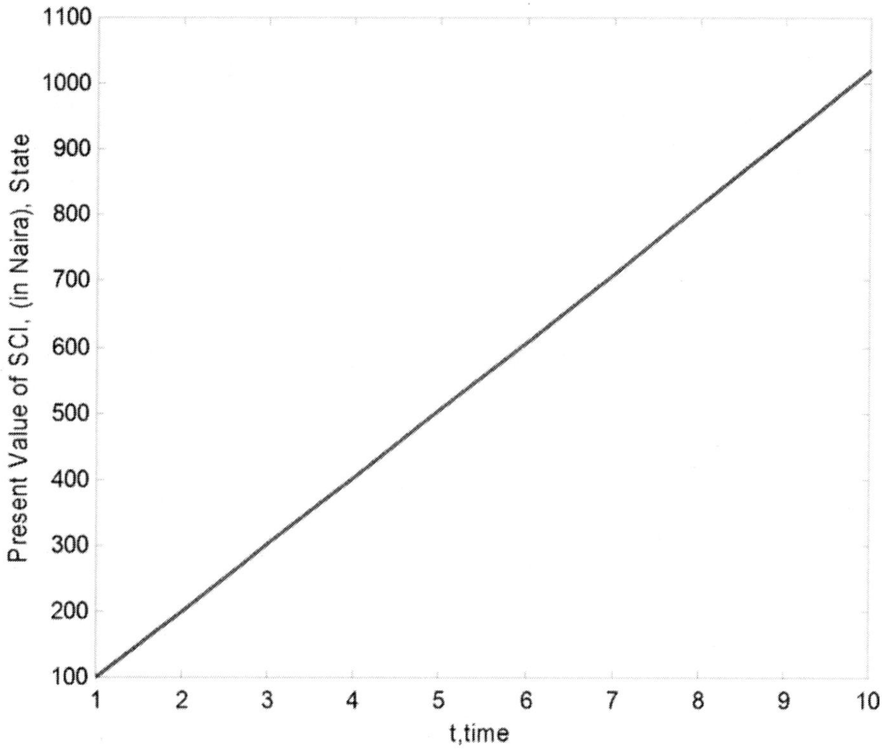

Figure 3. The present discounted value of future SCI.

$$\frac{\partial \Psi_0}{\partial r} = \frac{1}{\beta^2}\left(\left(1-\beta T\right)\frac{\partial \Psi_0}{\partial T} - D_0\right);$$

$$\frac{\partial \Psi_0}{\partial \sigma_D} = \theta \frac{\partial \Psi_0}{\partial r};$$

$$\frac{\partial \Psi_0}{\partial k} = \frac{1}{\beta^2}\left(\left(\beta T - 1\right)\frac{\partial \Psi_0}{\partial T} + D_0\right) = -\frac{\partial \Psi_0}{\partial r}.$$

Table 1 shows the sensitivity analysis of the discounted value of the SCI.

Proposition 3:

Suppose that Proposition 1 holds, then

$$d\Psi(t)=\Psi(t)\left(\left(r+\sigma_D\cdot\theta\right)dt+\sigma_D'\cdot dW(t)\right)$$
$$-D(t)dt. \tag{19}$$

Proof. Finding the differential of both sides of (16) and then substitute (7), we have

$$d\Psi(t)=\frac{D(t)\left(\exp\left(\beta(T-t)\right)-1\right)}{\beta}\times\left(kdt+\sigma_D'\cdot dW(t)\right)$$
$$+\frac{D(t)}{\beta}\left(-\beta\exp\left(\beta(T-t)\right)\right)dt$$
$$=\frac{D(t)}{\beta}\left(\left(\exp\left(\beta(T-t)\right)-1\right)\left(kdt+\sigma_D'\cdot dW(t)\right)\right.$$
$$\left.-\beta\left(\exp\left(\beta(T-t)\right)-1\right)dt\right)-D(t)dt$$
$$=\frac{D(t)\left(\exp\left(\beta(T-t)\right)-1\right)}{\beta}\left(\left(r+\sigma_D\cdot\theta\right)dt+\sigma_D'\cdot dW(t)\right)$$
$$-D(t)dt$$
$$=\Psi(t)\left(\left(r+\sigma_D\cdot\theta\right)dt+\sigma_D'\cdot dW(t)\right)-D(t)dt.$$

Let $V(t)$ be a value process at time t. We defined

Table 1. Simulation of the sensitivity analysis.

T	$\frac{\partial \Psi_0}{\partial D_0}$	$\frac{\partial \Psi_0}{\partial \sigma_1^D}$	$\frac{\partial \Psi_0}{\partial \sigma_2^D}$	$\frac{\partial \Psi_0}{\partial T}$	$\frac{\partial \Psi_0}{\partial r}$	$\frac{\partial \Psi_0}{\partial \theta_t}$	$\frac{\partial \Psi_0}{\partial \theta_s}$	$\frac{\partial \Psi_0}{\partial k}$
1	1.0022	−16624	−20005	0.4359	−207850	−51962	−74826	207850
2	2.0087	−14857	−17875	0.8738	−185720	−46429	−66858	185720
3	3.0197	−13079	−15736	1.3136	−163490	−40872	−58855	163490
4	4.0350	−11293	−13587	1.7552	−141160	−35290	−50818	141160
5	5.0548	−9439	−11429	2.1988	−118740	−29685	−42746	118740
6	6.0790	−7697	−9261	2.6444	−96220	−24054	−34638	96220
7	7.1077	−5838	−7084	3.0918	−73600	−18400	−26495	73600
8	8.1408	−4070	−4897	3.5413	−50880	−12720	−18317	50880
9	9.1785	−2245	−2701	3.9926	−28060	−7016	−10103	28060
10	10.2207	−412	−495	4.4460	−5150	−1287	−1853	5150

$V(t)$ as

$$V(t) := X(t) + \Psi(t), \tag{20}$$

where, $X(t)$ satisfy (10) and $\Psi(t)$ satisfy (19).

Proposition 4:

Let $V(t)$ satisfy (20), $X(t)$ satisfy (10) and $\Psi(t)$ satisfy (19), then

$$\begin{aligned} &dV(t) \\ &= \left(r\left(X(t)+\Psi\right)+\Psi(t)\sigma_D\cdot\theta+X(t)\lambda\cdot\Delta(t)\right)dt \\ &+\left(X(t)\Sigma\Delta(t)+\Psi(t)\sigma_D\right)'\cdot dW(t). \end{aligned} \tag{21}$$

Proof. Finding the differential of both sides of (20) and then substitute in (10) and (19), the result follows.

6. OPTIMIZATION OF THE VALUE OF WEALTH PROCESS

We define the general value function

$$J(t,v) = E\left[u\left(V\left(t, X(t), \Psi(t)\right)\right) \middle| X(t) = x, \Psi(t) = \Psi\right]$$

where $u(V(t))$ is the path of $V(t)$. Define Π to be the set of all admissible portfolio strategy that are F-progressively measurable, that satisfy the integrability conditions $E\left[\int_t^T \Delta'(u)\Delta(u)du\right] < \infty$ and let $U(V(t))$ be a concave function in $V(t)$ such that $U(V(t))$ satisfies the HJB equation

$$\begin{aligned} U(t,v) &= \sup_{\Delta\in\Pi} E\left[u(V(T)) \mid X(t)=x, \Psi(t)=\Psi\right] \\ &= \sup_{\Delta\in\Pi} J(t,v) \end{aligned}$$

Therefore, by applying Itô lemma on (21), we obtain the following HJB equation:

$$U_t(t,v) + \max_{\Delta\in\Pi} HV = 0, \tag{22}$$

subject to:

$$U(T,v) = \frac{v^{1-\gamma}}{1-\gamma},$$

where,

$$\begin{aligned} HV = {} & r\Psi U_\Psi + rxU_x + \Psi\sigma_D\cdot\theta U_\Psi + x\lambda\cdot\Delta(t)U_x \\ & + x\Psi\sigma_D\left(\Sigma\Delta(t)\right)U_{x\Psi} + \frac{1}{2}x^2\left(\Sigma\Delta(t)\right)'\Sigma\Delta(t)U_{xx} \\ & + \frac{1}{2}\Psi^2\sigma_D'\sigma_D U_{\Psi\Psi} \end{aligned} \tag{23}$$

Let $U(t,v)$ be the solution of the HJB equation (22). Since the utility function is concave and the value function is smooth, i.e., $U(t,V)\in C^{1,2}(\mathrm{R}\times[0,T])$, then (22) is well-defined. Hence, we have the following:

$$\frac{\partial HV}{\partial\Delta(t)}=\lambda U_x+x\left(\Sigma\Delta^*(t)\right)'\Sigma U_{xx}+\Psi\sigma_D\Sigma U_{X\Psi}$$
$$=0 \tag{24}$$

from (24), we have

$$\left(\Delta^*(t)\right)'=(\Sigma'\Sigma)^{-1}\left(\frac{-\lambda U_x}{xU_{xx}}\right)-\frac{(\Sigma')^{-1}\sigma_D\Psi U_{x\Psi}}{xU_{xx}}. \tag{25}$$

Substituting (25) into (22), we obtain the following HJB equation:

$$U_t+rxU_x+(r+\sigma_D\cdot\theta)\Psi U_\Psi+\frac{1}{2}\sigma_D'\sigma_D\Psi^2U_{\Psi\Psi}$$
$$-\frac{\lambda'\left(\Sigma^2\right)^{-1}\lambda U_x^2}{2U_{xx}}-\frac{\sigma_D'\sigma_D\Psi^2U_{x\Psi}^2}{2U_{xx}}$$
$$-\frac{\sigma_D'\left(\Sigma^{-1}\right)\lambda\Psi U_xU_{x\Psi}}{U_{xx}}=0. \tag{26}$$

This is the resulting HJB equation of our problem.

Proposition 5:

The solution of the HJB equation (26) is of the form

$$U(t,v)=\frac{(vA(t))^{1-\gamma}}{1-\gamma},\gamma\neq 0,\gamma>0,$$

with

$$A(t)=\exp\left(\left(r+\frac{\theta'\theta}{2\gamma}\right)(T-t)\right),$$
$$A(T)=1. \tag{27}$$

Proof. Finding the partial derivatives of

$U(t,v)=\frac{(vA(t))^{1-\gamma}}{1-\gamma}$ with respect to t, x, xx, Ψ,

$x\Psi$ and $\Psi\Psi$ and then substitute into (26), we obtain the following:

$$v^{1-\gamma}A(t)^{-\gamma}A'(t)+rv^{1-\gamma}A(t)^{1-\gamma}$$
$$+\frac{\lambda'(\Sigma^2)^{-1}\lambda v^{1-\gamma}A(t)^{1-\gamma}}{2\gamma}=0. \tag{28}$$

From (28), we obtain

$$A(t)=\exp\left(\left(r+\frac{\theta'\theta}{2\gamma}\right)(T-t)\right),$$
$$A(0)=1 \tag{29}$$

Using (29), we obtain

$$U(t,v)=\frac{v^{1-\gamma}}{1-\gamma}\exp\left(\left(r+\frac{\theta'\theta}{2\gamma}\right)(1-\gamma)(T-t)\right),$$
$$U(T,v)=\frac{v^{1-\gamma}}{1-\gamma}. \quad (30)$$

Figure 4 was obtained by

setting $D_0=100$, $r=0.04$, $k=0.099$, $T=10$, $\sigma_1^D=0.25$, $\sigma_2^D=0.36$, $\rho=0.6$, $\mu=0.09$, $\sigma_S=0.4$, $\sigma_I=0.3$ and $\theta_I=0.08$.

Figure 4 shows the expected value of utility of wealth at time $t\in[0,10]$ for different values of γ. Here, γ measures the level of risk the investor is willing to take. Observe that the smaller the value of γ, the higher the expected value of wealth, and vice versa. Therefore, when $\gamma=0.07$, the expected value of utility of wealth, $U(t,v)=U(10,v)=2780.9$, when $\gamma=0.08$, $U(10,v)=2271.5$, when $\gamma=0.09$, $U(10,v)=1914.1$ and when $\gamma=0.10$, $U(10,v)=1648.7$. Hence, the lower the rate at which investors are risk averse, the more the wealth that will accrue to them, and vice versa, which is an intuitive result.

Proposition 6:

Suppose that $U(t,V)$ is the solution of the HJB Equation (26), then the optimal portfolio in inflationlinked bond, stock and cash account are given by

$$\Delta_I^*(t)=\frac{\theta_I V^*(t)}{\gamma X^*(t)}-\frac{\sigma_1^D\Psi(t)}{\sigma_I X^*(t)}. \quad (31)$$

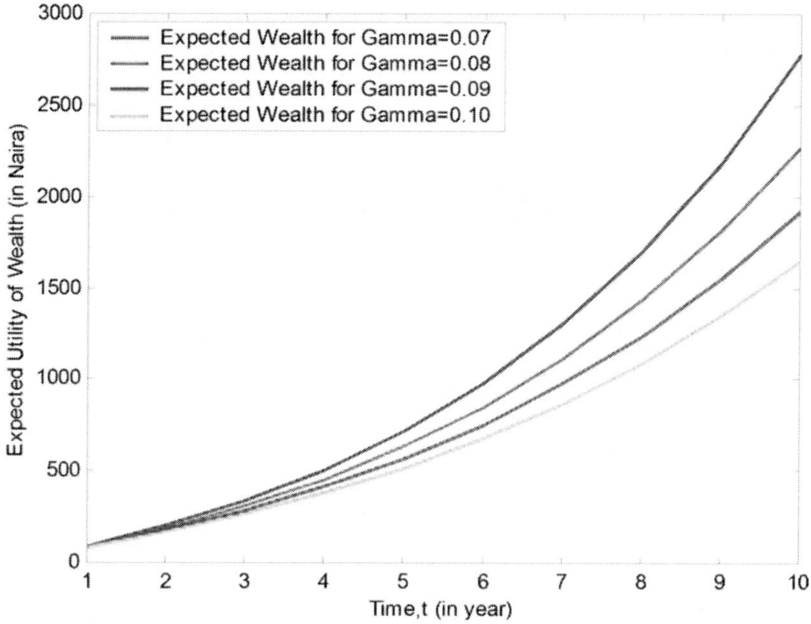

Figure 4. The expected value of utility of wealth for different values of γ.

$$\Delta_S^*(t) = \frac{\left(\sigma_I(\mu - r) + \theta_I \sigma_S \rho\left(\sigma_S\sqrt{1-\rho^2} - \sigma_I\right)\right)V^*(t)}{\gamma\left(1-\rho^2\right)\sigma_I \sigma_S^2 X^*(t)} - \frac{\left(\sigma_I \sigma_2^D - \sigma_S \sigma_1^D \rho\right)\Psi(t)}{\sigma_I \sigma_S \sqrt{1-\rho^2}\, X^*(t)}. \quad (32)$$

$$\Delta_0^*(t) = 1 - \frac{V^*(t)}{X^*(t)}\left(\frac{\theta_I}{\gamma} + \frac{\left(\sigma_I(\mu - r) + \theta_I \sigma_S \rho\left(\sigma_S\sqrt{1-\rho^2} - \sigma_I\right)\right)}{\gamma\left(1-\rho^2\right)\sigma_I \sigma_S^2}\right) + \frac{\Psi(t)}{X^*(t)}\left(\frac{\sigma_1^D}{\sigma_I} + \frac{\left(\sigma_I \sigma_2^D - \sigma_S \sigma_1^D \rho\right)}{\sigma_I \sigma_S \sqrt{1-\rho^2}}\right).$$

Proof. From (25), we have that

$$\Delta^*(t) = \frac{(\Sigma^2)^{-1} \lambda V^*(t)}{\gamma X^*(t)} - \frac{\Sigma^{-1}\sigma_D \Psi(t)}{X^*(t)}.$$

But,

$$\Sigma^{-1}\sigma_D = \begin{pmatrix} \dfrac{\sigma_1^D}{\sigma_I} \\ \dfrac{\sigma_I \sigma_2^D - \sigma_S \sigma_1^D \rho}{\sigma_S \sigma_I \sqrt{1-\rho^2}} \end{pmatrix}$$

$$\Sigma^{-2}\lambda = \begin{pmatrix} \dfrac{\theta_I}{\sigma_I} \\ \dfrac{\sigma_I(\mu - r) + \sigma_S \theta_I \rho \left(\sigma_S \sqrt{1-\rho^2} - \sigma_I\right)}{\sigma_S^2 \sigma_I (1-\rho^2)} \end{pmatrix}$$

Therefore,

$$\Delta^*(t)=\begin{pmatrix}\Delta_I^*(t)\\ \Delta_S^*(t)\end{pmatrix}=\frac{V^*(t)}{\gamma X^*(t)}\begin{pmatrix}\dfrac{\theta_I}{\sigma_I}\\ \dfrac{\sigma_I(\mu-r)+\sigma_S\theta_I\rho\left(\sigma_S\sqrt{1-\rho^2}-\sigma_I\right)}{\sigma_S^2\sigma_I(1-\rho^2)}\end{pmatrix}-\frac{\Psi(t)}{X^*(t)}\begin{pmatrix}\dfrac{\sigma_1^D}{\sigma_I}\\ \dfrac{\sigma_I\sigma_2^D-\sigma_S\sigma_1^D\rho}{\sigma_S\sigma_I\sqrt{1-\rho^2}}\end{pmatrix}$$

$$=\begin{pmatrix}\dfrac{\theta_I V^*(t)}{\sigma_I\gamma X^*(t)}\\ \dfrac{\left(\sigma_I(\mu-r)+\sigma_S\theta_I\rho\left(\sigma_S\sqrt{1-\rho^2}-\sigma_I\right)\right)V^*(t)}{\sigma_S^2\sigma_I(1-\rho^2)\gamma X^*(t)}\end{pmatrix}-\begin{pmatrix}\dfrac{\sigma_1^D\Psi(t)}{\sigma_I X^*(t)}\\ \dfrac{\left(\sigma_I\sigma_2^D-\sigma_S\sigma_1^D\rho\right)\Psi(t)}{\sigma_S\sigma_I\sqrt{1-\rho^2}X^*(t)}\end{pmatrix}$$

$$=\begin{pmatrix}\dfrac{\theta_I V^*(t)}{\sigma_I\gamma X^*(t)}-\dfrac{\sigma_1^D\Psi(t)}{\sigma_I X^*(t)}\\ \dfrac{\left(\sigma_I(\mu-r)+\sigma_S\theta_I\rho\left(\sigma_S\sqrt{1-\rho^2}-\sigma_I\right)\right)V^*(t)}{\sigma_S^2\sigma_I(1-\rho^2)\gamma X^*(t)}-\dfrac{\left(\sigma_I\sigma_2^D-\sigma_S\sigma_1^D\rho\right)\Psi(t)}{\sigma_S\sigma_I\sqrt{1-\rho^2}X^*(t)}\end{pmatrix}$$

It implies that

$$\Delta_I^*(t)=\frac{\theta_I V^*(t)}{\gamma X^*(t)}-\frac{\sigma_1^D\Psi(t)}{\sigma_I X^*(t)}. \tag{33}$$

$$\begin{aligned}&\Delta_S^*(t)\\ &=\frac{\left(\sigma_I(\mu-r)+\theta_I\sigma_S\rho\left(\sigma_S\sqrt{1-\rho^2}-\sigma_I\right)\right)V^*(t)}{\gamma(1-\rho^2)\sigma_I\sigma_S^2X^*(t)}\\ &-\frac{\left(\sigma_I\sigma_2^D-\sigma_S\sigma_1^D\rho\right)\Psi(t)}{\sigma_I\sigma_S\sqrt{1-\rho^2}X^*(t)}.\end{aligned} \tag{34}$$

$$\Delta_0^*(t) = 1 - \left(\frac{\theta_I V^*(t)}{\gamma X^*(t)} + \frac{\left(\sigma_I(\mu - r) + \theta_I \sigma_S \rho\left(\sigma_S\sqrt{1-\rho^2} - \sigma_I\right)\right)V^*(t)}{\gamma\left(1-\rho^2\right)\sigma_I \sigma_S^2 X^*(t)} \right) + \frac{\sigma_1^D \Psi(t)}{\sigma_I X^*(t)} + \frac{\left(\sigma_I \sigma_2^D - \sigma_S \sigma_1^D \rho\right)\Psi(t)}{\sigma_I \sigma_S \sqrt{1-\rho^2}\, X^*(t)}. \tag{35}$$

From (33), the first term represents the classical portfolio strategy while the second term represents the intertemporal hedging strategy that offset any shock to the SCI at time t. From (34), $\frac{\left(\sigma_I(\mu - r) + \theta_I \sigma_S \rho\left(\sigma_S\sqrt{1-\rho^2} - \sigma_I\right)\right)V^*(t)}{\gamma\left(1-\rho^2\right)\sigma_I \sigma_S^2 X^*(t)}$ represents the classical portfolio strategy at time t and $\frac{\left(\sigma_I \sigma_2^D - \sigma_S \sigma_1^D \rho\right)\Psi(t)}{\sigma_I \sigma_S \sqrt{1-\rho^2}\, X^*(t)}$ represents the intertemporal hedging term that offset shock resulting from the SCI at time t. From (35), observe that these hedging terms can be transfer to cash account at time t or it can be reinvest in stock and in inflation-linked bond at time t.

Figure 5 was obtained by setting $D_0 = 100$, $r = 0.04$, $k = 0.099$, $T = 10$, $\sigma_1^D = 0.25$, $\sigma_2^D = 0.36$, $\rho = 0.6$, $\mu = 0.09$, $\sigma_S = 0.4$, $\sigma_I = 0.3$, $\theta_I = 0.08$ and $\gamma = 0.5$. The optimal portfolio value in inflation-linked bond at time $t = 10$ is obtained as 0.16 (or 16%).

Figure 6 was obtained by setting $D_0 = 100$, $r = 0.04$, $k = 0.099$, $T = 10$, $\sigma_1^D = 0.25$, $\sigma_2^D = 0.36$, $\rho = 0.6$, $\mu = 0.09$, $\sigma_S = 0.4$,

$\sigma_I = 0.3$, $\theta_I = 0.08$ and $\gamma = 0.5$. The optimal portfolio value in stock at time $t=10$ is obtained as 1.001562 (or 100.1562%).

Figure 7 was obtained by setting $D_0 = 100$, $r = 0.04$, $k = 0.099$, $T = 10$, $\sigma_1^D = 0.25$, $\sigma_2^D = 0.36$, $\rho = 0.6$, $\mu = 0.09$, $\sigma_S = 0.4$, $\sigma_I = 0.3$, $\theta_I = 0.08$ and $\gamma = 0.5$. The optimal portfolio value in cash account at time $t = 10$ is obtained as −0.161562 (or −16.1562%).

In Figures 5-7, we set $D_0 = 100$, $r = 0.04$, $k = 0.099$, $T = 10$, $\sigma_1^D = 0.25$, $\sigma_2^D = 0.36$, $\rho = 0.6$, $\mu = 0.09$, $\sigma_S = 0.4$, $\sigma_I = 0.3$, $\theta_I = 0.08$ and $\gamma = 0.10$. **Figure 5** shows the portfolio value in inflation-linked bond. We found that the optimal portfolio value in inflation-linked bond at time $t=10$ is 0.16 (or 16%). **Figure 6** shows the portfolio value in stock. We found that the optimal portfolio value in stock at time $t=10$ is 1.001562 (or 100.1562%). **Figure 7** shows the portfolio value in cash account. We found that the optimal portfolio value in cash account at time t = 10 is −0.161562 (or −16.1562%).

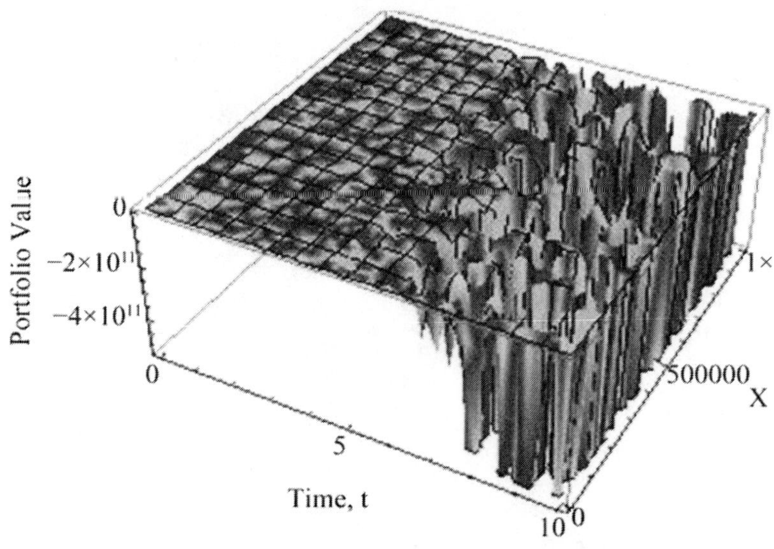

Figure 5. Portfolio value in inflation-linked bond.

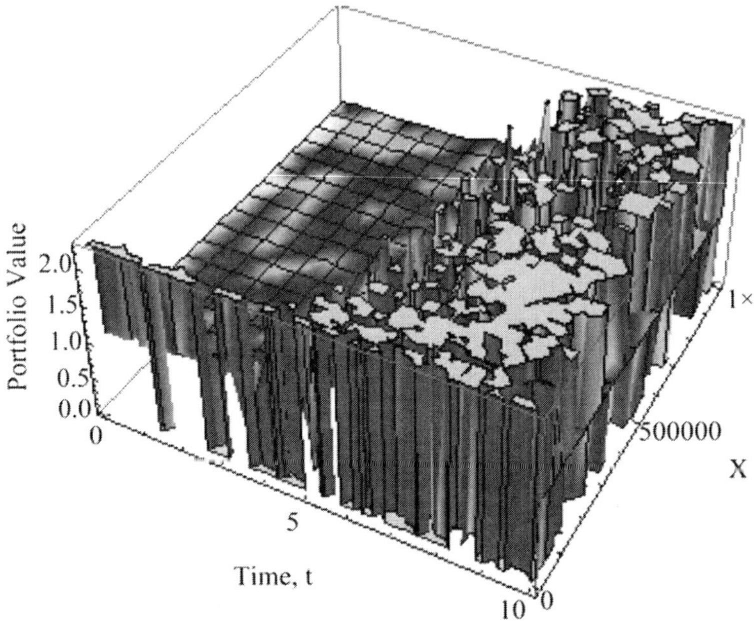

Figure 6. Portfolio value in stock.

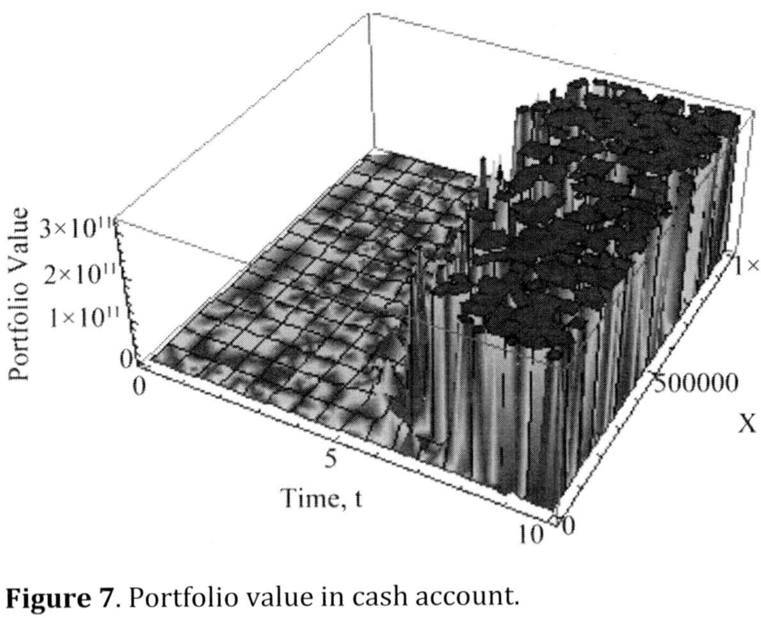

Figure 7. Portfolio value in cash account.

7. CONCLUSION

The optimal portfolio strategy X with discounted stochastic cash inflows was considered. It was assumed that the cash inflow, stock and inflation-linked bond are stochastic and follow a standard geometric Brownian motion. The sensitivity analysis of the present value of the discounted cash inflows was carried out in this paper and the results are presented in **Table 1**. Analytical solution to the resulting HJB equation was obtained. It was found that the smaller the value of γ (which measure the level of risk the investor is willing to take), the higher the expected value of wealth, and vice versa. The optimal portfolio values in stock, inflation-linked bond and cash account were obtained. The resulting optimal portfolio values in stock and inflation-linked bond were found to involve intertemporal hedging terms that offset any shock to the SCI.

REFERENCES

1. G. Deelstra, M. Grasselli and P. Koehl, "Optimal Investment Strategies in a CIR Framework," Journal of Applied Probability, Vol. 37, No. 4, 2000, pp. 936-946.

2. D. O. Cajueiro and T. Yoneyama, "Optimal Portfolio, Optimal Consumption and the Markowitz Mean-Variance Analysis in a Switching Diffusion Market," 2003. unb.br/face/eco/seminarios/sem0803.pdf

3. A. Zaks, "Present Value of Annuities under Random Rates of Interest," 2003. http://academic.research.microsoft.com/Publication/6244175/present-value-of-annuities-under-random-rates-of-interest

4. D. Dentcheva and A. Ruszczynski, "Portfolio Optimization with Stochastic Dominance Constraints," SIAM Journal on Optimization, Vol. 14, No. 2, 2003, pp. 548- 566.

5. D. Blake, "Efficiency, Risk Aversion and Portfolio Insurance: An Analysis of Financial Asset Portfolios Held by Investors in the United Kindom," Economic Journal, Vol. 106, No. 438, 1996, pp. 1175-1192.

6. A. Zhang, "Stochastic Optimization in Finance and Life Insurance: Applications of the Martingale Method," Ph.D. Thesis, University of Kaiserslauten, Kaiserslautern, 2007.

7. J. Mukuddem-Peterson, M. A. Peterson and I. M. Schoeman, "An Application of Stochastic Optimization Theory to Institutional Finance," Applied Mathematics Sciences, Vol. 1, No. 28, 2007, pp. 1359-1385.

8. P. Battocchio, "Optimal Portfolio Strategies with Stochastic Wage Income: The Case of a Defined Contribution Pension Plan," Working Paper, Université Catholique de Louvain, Louvain-la-Neuve, 2002.

9. B. H. Lim and U. J. Choi, "Optimal Consumption and Portfolio Selection with Portfolio Constraints," International Sciences, Vol. 4, 2009, pp. 293-309.

10. C. I. Nkeki, "On Optimal Portfolio Management of the Accumulation Phase of a Defined Contributory Pension Scheme," Ph.D. Thesis, University of Ibadan, Ibadan, 2011.

11. C. I. Nkeki and C. R. Nwozo, "Variational Form of Classical Portfolio Strategy and Expected Wealth for a Defined Contributory Pension Scheme," Journal of Mathematical Finance, Vol. 2, No. 1, 2012, pp. 132-139.

4

CHAPTER

Stochastic Programming Model for Discrete Lotsizing and Scheduling Problem on Parallel Machines

Kensuke Ishiwata[1], Jun Imaizumi[2], Takayuki Shiina[3], Susumu Morito[4]

[1]*Global Logistics, Kao Corporation, Tokyo, Japan*

[2]*Faculty of Business Administration, Toyo University, Tokyo, Japan*

[3]*Faculty of Social Systems Science, Chiba Institute of Technology, Chiba, Japan*

[4]*School of Creative Science and Engineering, Waseda University, Tokyo, Japan*

ABSTRACT

In recent years, it has been difficult for manufactures and suppliers to forecast demand from a market for a given product precisely. Therefore, it has become important for them to cope with fluctuations in demand. From this viewpoint, the problem of planning or scheduling in production systems can be regarded as a mathematical problem with stochastic elements. However, in many previous studies, such problems are formulated without stochastic factors, treating stochastic elements as deterministic variables or parameters. Stochastic programming incorporates such factors into the mathematical formulation. In the present paper, we consider a multi-product, discrete, lotsizing and scheduling problem on parallel machines with stochastic demands. Under certain assumptions, this problem can be formulated as a stochastic integer programming problem. We attempt to solve this problem by a scenario aggregation method proposed by Rockafellar and Wets. The results from computational experiments suggest that our approach is able to solve large-scale problems, and that, under the condition of uncertainty, incorporating stochastic elements into the model gives better results than formulating the problem as a deterministic model.

Keywords: Stochastic Programming; Lotsizing and Scheduling; Parallel Machines; Scenario Aggregation Method

1. STUDY BACKGROUND AND OBJECTIVE

In the manufacturing and supply chain industries in recent times, one of the most important challenges has been to determine how best to deal with the significant fluctuations in demand. Within the market environment of today, it is difficult to forecast customer behavior and not uncommon for any forecasts of demand that are made to miss the mark. As such, it has become necessary to take into account the element of uncertainty when developing plans. However, a challenge faced in production planning is the lot-scheduling problem. This stems from the fact that the main focus in production planning is on determination of lot sizes, which corresponds to tacticallevel decision-making, and determination of scheduling, which corresponds to operational-level decision-making, where the goal is to optimize both of these factors concurrently.

Almost all past research into the lot-scheduling problem has considered demand as a definite value. However, in light of the above-mentioned circumstances, it seems that in many cases, demand should be regarded as an uncertain element. For example, even if one is able to treat the most recent demand with certainty, it is quite likely that when decisions are made at a point in time later, uncertain elements will come into the mix. In such a situation, it becomes necessary to develop a plan which also takes into account those uncertain elements. Mathematical programming which takes these probabilistic factors into consideration is known as stochastic programming (Birge [1], Birge and Louveaux [2], Shiina [3]).

In the present study, we looked at a lot-scheduling problem on parallel machines and considered a stochastic programming model. Compared with a deterministic programming model that does not incorporate probabilistic factors, the stochastic programming model poses a large-scale problem for which it is often extremely difficult to find a solution. With this in mind, we felt it would be possible to obtain a solution by devising an approximate solution method based on the scenario aggregation method proposed by Rockafellar and Wets [4]. Specifically, we referred to the method of Løkketangen and Woodruff [5], which presents a specific general framework for problems incorporating 0 - 1 variables, to develop a procedure which provides an approximate solution.

We investigate the advantages of the stochastic programming model by numerical simulation to first compare the results obtained when following a deterministic programming model with the results obtained when the problem is considered in terms of a stochastic programming model, and we then assess the performance of the procedure. Next, we discuss the accuracy of the approximate solution and the computation time, and demonstrate the effectiveness of the solution method we developed.

2. LOT-SCHEDULING PROBLEM

2.1. Definition of Problem

The lot-scheduling problem is to decide both the production lot size for each item type and the production time for each lot, in a lot production process in which preparatory work is required. In the present study, we consider a lot-scheduling problem on parallel machines. A summary of the problem is as follows.

Production is performed in which multiple types of items are exchanged around on multiple machines with differing performance that make up a single process. When the item types are exchanged, time and expense are required for preparatory work. Inventory storage costs arise in proportion to the amount of inventory. For each item type, an ordered shipping amount (demand quantity) is provided at designated times, and no item is allowed to run out. At any one time, a machine is capable of producing only one item type. The machines can be used only for a certain duration at a time, and it is necessary to carry out both preparation and production during that period. If the unit manufacturing cost is to be lowered and the lot size increased, then the expense incurred for storage of inventory increases; conversely, if the inventory is reduced, then preparatory work will need to be carried out more frequently, leading to the problem of increased time and expense for preparation.

2.2. Past Research and the Relevance of the Present Study

Past research into the lot-scheduling problem on parallel machines has, in many cases, studied deterministic programming models which treat customer demand as a definite value (for example, Meyr [6], Arai et al. [7]). So far, there has been almost no research into stochastic programming models for the lot-scheduling problem on parallel machines. Therefore, we assumed a situation in which some of the demand had to be treated as uncertain and formulated the lot-scheduling problem on parallel machines in terms of a stochastic programming model and then calculated a solution by applying the scenario aggregation method.

3. FORMULATION

3.1. Deterministic Programming Model

Item type, machine, and time are represented by N, M, and T, respectively. For the constants, the demand for item type n at time t is d_{nt}, the length of time taken to create one unit of item type n on machine m is p_{nm}, the length of time that machine m can be used at time t is L_{mt}, the cost of preparation for item type n on machine m is c_{nm}, the time taken for preparation is C_{nm}, and the cost of storage of inventory for item type n is h_n. For item type n on machine m at time t, the variable representing the production amount is x_{nmt}, the variable indicating whether production is occurring is y_{nmt}, and the variable indicating whether preparation is occurring is z_{nmt}. The variable representing the amount of inventory for item type n at time t is I_{nt}. The formulas used are as follows.

$$\min \sum_{n,m,t} c_{nm} z_{nmt} + \sum_{n,t} h_n I_{nt} \quad (1)$$

$$\text{s.t.} I_{nt} = I_{n(t-1)} + \sum_m x_{nmt} - d_{nt}, \forall n,t \quad (2)$$

$$p_{nm} x_{nmt} + C_{nm} z_{nmt} \leq L_{mt} y_{nmt}, \forall n,m,t \quad (3)$$

$$\sum_n y_{nmt} \leq 1, \forall m,t \quad (4)$$

$$z_{nmt} \geq y_{nmt} - y_{nm(t-1)}, \forall n,m,t \quad (5)$$

$$y_{nm0} = 0, \forall n,m \quad (6)$$

$$I_{n0} = 0, \forall n \quad (7)$$

$$x_{nmt} \geq 0, \forall n,m,t \quad (8)$$

$$y_{nmt} \in \{0,1\}, \forall n,m,t \tag{9}$$

$$z_{nmt} \geq 0, \forall n,m,t \tag{10}$$

$$I_{nt} \geq 0, \forall n,t \tag{11}$$

Here, (1) shows the objective function: the sum total of the preparation costs and the inventory storage costs; (2) updates the changes in inventory; (3) is the constraint on the time that a machine can be used; (4) reflects that, at each time, a machine is only capable of producing one type of item; (5) is a constraint on preparation being carried out; (6) and (7) are the initial conditions for production and inventory; and (8), (9), (10), and (11) are additional conditions on the individual variables.

3.2. Changes in Demand

In this section, we will discuss demand d, omitting the subscripted symbol for the item type. Specifically, this means that even though the item type is not indicated, the discussion assumes a specific item type.

Demand at time t, $\tilde{d}_t$, is defined as a stochastic variable, designated d_t, which is assumed to follow a finite discrete distribution. The list of possible values for the stochastic variable across time T, represented $d = (d_1, \cdots, d_T)$, is referred to as a scenario.

Based on the discrete and finite properties of the distribution, the number of scenarios is represented with the symbol S. The probability that a scenario s, that is to say $d^s = (d_1^s, \cdots, d_T^s)$, will occur is given by P^s (where $\sum_{s=1}^{S} P^s = 1$). In the scenario tree (the directed graph in **Figure 1**), the scenario is represented as routes from the root node to the leaf nodes.

If the possible values for demand d^{s_1} and d^{s_2} of two scenarios s_1 and s_2 (with ($s_1 \neq s_2$)) satisfy the equality $(d_1^{s_1}, \cdots, d_t^{s_1}) = (d_1^{s_2}, \cdots, d_t^{s_2})$ over the history to a certain time t, then they follow the same route on the tree, to time t.

The decision making for the two scenarios s_1 and s_2 must be the same to time t. At time t, the individual making decisions cannot anticipate that scenarios s_1 and s_2 will diverge in the future. This is because, at time t, they are not provided with information regarding the future development from time t + 1 onward, and they have to make a decision based on the history of dt until time t. This condition is referred to as the nonanticipativity condition.

The set of scenarios (represented as numbers) $\{1, \cdots, S\}$ can be partitioned into disjoint subsets at each time. The index set for scenarios equal to scenario s in history to time t is represented by $B(s,t)$ and referred to as the scenario bundle. For example, in **Figure 1**, B(1, 1) = B(2, 1) = {1, 2}, and B(3, 1) = B(4, 1) = {3, 4}.

3.3. Stochastic Programming Model

The amount of demand for item type n at time t in scenario s is represented by d_{nt}^{s} and the probability of occurrence of scenario s is P^s, as described above. Furthermore, the variable representing the production amount for item type n by machine m at time t in scenario s is x_{nmt}^{s}, the variable indicating whether production is being carried out is y_{nmt}^{s}, and the variable indicating whether preparation is being carried out is z_{nmt}^{s}. The variable representing the amount of inventory of item type n at time t in scenario s is I_{nt}^{s}.

The remaining symbols are as defined in Section 3.1. The relevant formulation is as follows.

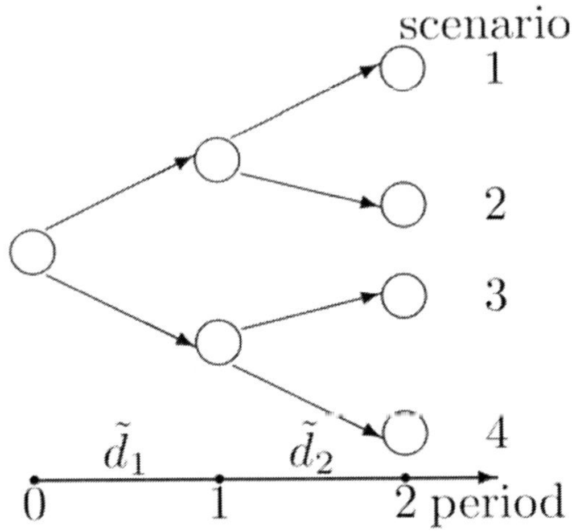

Figure 1. Scenario tree.

$$\min \sum_{s} P^s \left(\sum_{n,m,t} c_{nm} z^s_{nmt} + \sum_{n,t} h_n I^s_{nt} \right) \tag{12}$$

$$\text{s.t.} I^s_{nt} = I^s_{n(t-1)} + \sum_{m} x^s_{nmt} - d^s_{nt}, \forall s,n,t \tag{13}$$

$$p_{nm} x^s_{nmt} + C_{nm} z^s_{nmt} \leq L_{mt} y^s_{nmt}, \forall s,n,m,t \tag{14}$$

$$\sum_{n} y^s_{nmt} \leq 1, \forall s,m,t \tag{15}$$

$$z^s_{nmt} \geq y^s_{nmt} - y^s_{nm(t-1)}, \forall s,n,m,t \tag{16}$$

$$y_{nm0}^{s} = 0, \forall s, n, m \tag{17}$$

$$I_{n0}^{s} = 0, \forall s, n \tag{18}$$

$$x_{nmt}^{s_1} = x_{nmt}^{s_2}, \forall s_1, s_2, n, m, t, s_1 \neq s_2, B(s_1, t) = B(s_2, t) \tag{19}$$

$$y_{nmt}^{s_1} = y_{nmt}^{s_2}, \forall s_1, s_2, n, m, t, s_1 \neq s_2, B(s_1, t) = B(s_2, t) \tag{20}$$

$$z_{nmt}^{s_1} = z_{nmt}^{s_2}, \forall s_1, s_2, n, m, t, s_1 \neq s_2, B(s_1, t) = B(s_2, t) \tag{21}$$

$$I_{nt}^{s_1} = I_{nt}^{s_2}, \forall s_1, s_2, n, t, s_1 \neq s_2, B(s_1, t) = B(s_2, t) \tag{22}$$

$$x_{nmt}^{s} \geq 0, \forall s, n, m, t \tag{23}$$

$$y_{nmt}^{s} \in \{0,1\}, \forall s, n, m, t \tag{24}$$

$$z_{nmt}^{s} \geq 0, \forall s, n, m, t \tag{25}$$

$$I^{s}_{nt} \geq 0, \forall s, n, t \tag{26}$$

Here, (12) shows the objective function: the expectation of the sum total of the preparation costs and inventory storage costs; (13) updates the changes in inventory; (14)is the constraint on the time that a machine can be used; (15) reflects that, at each time, a machine is only capable of producing one type of item; (16) is a constraint on preparation being carried out; (17) and (18) are the initial conditions for production and inventory; (19), (20), (21), and (22) are the nonanticipativity conditions; and (23), (24), (25), and (26) are additional conditions on the individual variables.

4. APPLYING THE SCENARIO AGGREGATION METHOD

4.1. Equivalent Deterministic Programming Problems

The formulation of problems (not only the lot-scheduling problem on parallel machines) via a stochastic programming model such as this poses a large-scale mathematical programming challenge, due to the variables that must be defined and the constraint conditions that must be listed for all of the scenarios. The lot-scheduling problem, which is modeled by the stochastic programming model considered in the present study, in particular, involves very substantial difficulties, if it is to be solved in its original format, as a mixed-integer programming problem. Therefore, in the present study, we took the problem structure arising from the modeling and formulation into account, and opted to pursue an approximate solution by using a framework known as the scenario aggregation method, which was proposed by Rockafellar and Wets [4].

4.2. Overview of the Scenario Aggregation Method

The scenario aggregation method is a solution method wherein a scenario sub-problem is defined for each route on the scenario tree leading from the root to a leaf, and the solutions obtained by solving the scenario sub-problems are aggregated. An overview of this solution method follows.

First, solutions are computed by solving the scenario sub-problems (we call these the admissible solutions). Next, the allowable solutions are aggregated, in order to compute a solution that satisfies the nonanticipativity condition (we call this the implementable solution).

The admissible solutions are guaranteed to be practicable for each scenario, but do not necessarily satisfy the nonanticipativity condition. In contrast, the implementable solution is guaranteed to satisfy the nonanticipativity condition, but is not guaranteed to be feasible for each scenario. Therefore, in the scenario aggregation method, the difference between these two types of solution is added to the scenario sub-problem in the form of a quadratic penalty term, and these differences are gradually allowed to converge until finally a solution to the problem to be solved can be computed.

The following section describes this process in detail.

4.3. Solution Algorithm

Below is described the specific procedure. In order to reduce the computation time required to reach the end condition in the scenario aggregation method, following the research of Løkketangen and Woodruff [5], we took into consideration only the convergence of integer variables, not the convergence of continuous variables.

Step 1

The following initial scenario sub-problem, corresponding to a scenario s, is solved and the admissible solutions $x^s_{nmt}, y^s_{nmt}, z^s_{nmt}, I^s_{nt}$ are computed.

$$\min \sum_{n,m,t} c_{nm} z^s_{nmt} + \sum_{n,t} h_n I^s_{nt}$$

$$\text{s.t. } I^s_{nt} = I^s_{n(t-1)} + \sum_m x^s_{nmt} - d^s_{nt}, \forall n,t$$

$$p_{nm} x^s_{nmt} + C_{nm} z^s_{nmt} \le L_{mt} y^s_{nmt}, \forall n,m,t$$

$$\sum_n y^s_{nmt} \le 1, \forall m,t$$

$$z^s_{nmt} \ge y^s_{nmt} - y^s_{nm(t-1)}, \forall n,m,t$$

$$y^s_{nm0} = 0, \forall n,m$$

$$I^s_{n0} = 0, \forall n,$$

$$x^s_{nmt} \ge 0, \forall n,m,t$$

$$y^s_{nmt} \in \{0,1\}, \forall n,m,t$$

$$z_{nmt}^{s} \geq 0, \forall n, m, t$$

$$I_{nt}^{s} \geq 0, \forall n, t \; I_{nt}^{s} \geq 0, \forall n, t$$

Step 2

The following aggregation is applied to the admissible solutions, and the implementable solutions $\hat{x}_{nmt}^{s}, \hat{y}_{nmt}^{s}, \hat{z}_{nmt}^{s}, \hat{I}_{nt}^{s}$ are computed. The values of the Lagrange multipliers $\alpha_{nmt}^{s}, \beta_{nmt}^{s}, \gamma_{nmt}^{s}, \delta_{nt}^{s}$ are set to 0 and we go to Step 3.

$$\hat{x}_{nmt}^{s} := \sum_{s \in B(s,t)} P^{s} x_{nmt}^{s} \div \sum_{s \in B(s,t)} P^{s}$$

$$\hat{y}_{nmt}^{s} := \sum_{s \in B(s,t)} P^{s} y_{nmt}^{s} \div \sum_{s \in B(s,t)} P^{s}$$

$$\hat{z}_{nmt}^{s} := \sum_{s \in B(s,t)} P^{s} z_{nmt}^{s} \div \sum_{s \in B(s,t)} P^{s}$$

$$\hat{I}_{nt}^{s} := \sum_{s \in B(s,t)} P^{s} I_{nt}^{s} \div \sum_{s \in B(s,t)} P^{s}$$

Step 3

A scenario sub-problem that has the following 0-1 constraints, corresponding with each scenario s, is solved and the admissible solutions x^s_{nmt}, y^s_{nmt}, z^s_{nmt}, I^s_{nt} are computed. The objective function terms starting from the third term are penalty terms for the admissible and implementable solutions; ρ is a parameter in the scenario aggregation method.

$$\min \sum_{n,m,t} c_{nm} z^s_{nmt} + \sum_{n,t} h_n I^s_{nt}$$

$$+\sum_{n,m,t} \left\{ \alpha^s_{nmt} x^s_{nmt} + \frac{1}{2}\rho\left(x^s_{nmt} - \hat{x}^s_{nmt}\right)^2 \right\}$$

$$+\sum_{n,m,t} \left\{ \beta^s_{nmt} y^s_{nmt} + \frac{1}{2}\rho\left(y^s_{nmt} - \hat{y}^s_{nmt}\right)^2 \right\}$$

$$+\sum_{n,m,t} \left\{ \gamma^s_{nmt} z^s_{nmt} + \frac{1}{2}\rho\left(z^s_{nmt} - \hat{z}^s_{nmt}\right)^2 \right\}$$

$$+\sum_{n,t} \left\{ \delta^s_{nt} I^s_{nt} + \frac{1}{2}\rho\left(I^s_{nt} - \hat{I}^s_{nt}\right)^2 \right\}$$

$$\text{s.t. } I^s_{nt} = I^s_{n(t-1)} + \sum_m x^s_{nmt} - d^s_{nt}, \forall n,t$$

$$p_{nm} x^s_{nmt} + C_{nm} z^s_{nmt} \le L_{mt} y^s_{nmt}, \forall n,m,t$$

$$\sum_n y^s_{nmt} \le 1, \forall m,t$$

$$z^s_{nmt} \ge y^s_{nmt} - y^s_{nm(t-1)}, \forall n,m,t$$

$$y^{s}_{nm0} = 0, \forall n, m$$

$$I^{s}_{n0} = 0, \forall n$$

$$x^{s}_{nmt} \geq 0, \forall n, m, t$$

$$y^{s}_{nmt} \in \{0,1\}, \forall n, m, t$$

$$z^{s}_{nmt} \geq 0, \forall n, m, t$$

$$I^{s}_{nt} \geq 0, \forall n, t$$

Step 4

Aggregation is applied to the admissible solutions, and the implementable solutions $\hat{x}^{s}_{nmt}$, $\hat{y}^{s}_{nmt}$, $\hat{z}^{s}_{nmt}$, $\hat{I}^{s}_{nt}$ are computed. If the admissible solution y^{s}_{nmt} and the implementable solution $\hat{y}^{s}_{nmt}$ become equal, we go to Step 5. Otherwise, the values of the Lagrange multipliers α^{s}_{nmt}, β^{s}_{nmt}, γ^{s}_{nmt}, δ^{s}_{nt} are altered using the equations below and we go to Step 3.

$$\alpha^{s}_{nmt} := \alpha^{s}_{nmt} + \rho\left(x^{s}_{nmt} - \hat{x}^{s}_{nmt}\right)$$

$$\beta^s_{nmt} := \beta^s_{nmt} + \rho\left(y^s_{nmt} - \hat{y}^s_{nmt}\right)$$

$$\gamma^s_{nmt} := \gamma^s_{nmt} + \rho\left(z^s_{nmt} - \hat{z}^s_{nmt}\right)$$

$$\delta^s_{nt} := \delta^s_{nt} + \rho\left(I^s_{nt} - \hat{I}^s_{nt}\right)$$

Step 5

The value of the converged integer variable $\overline{y}^s_{nmt}$ is substituted into the original problem formulation and the following linear programming problem is solved. The values of the continuous variables x^s_{nmt}, z^s_{nmt}, I^s_{nt} are computed to complete the procedure.

$$\min \sum_s P^s \left(\sum_{n,m,t} c_{nm} z^s_{nmt} + \sum_{n,t} h_n I^s_{nt} \right)$$

$$\text{s.t. } I^s_{nt} = I^s_{n(t-1)} + \sum_m x^s_{nmt} - d^s_{nt}, \forall s,n,t$$

$$p_{nm} x^s_{nmt} + C_{nm} z^s_{nmt} \leq L_{mt} \overline{y}^s_{nmt}, \forall s,n,m,t$$

$$z^s_{nmt} \geq \overline{y}^s_{nmt} - \overline{y}^s_{nm(t-1)}, \forall s,n,m,t$$

$$I_{n0}^{s} = 0, \forall s, n, m$$

$$x_{nmt}^{s_1} = x_{nmt}^{s_2}, \forall s_1, s_2, n, m, t,$$

$$s_1 \neq s_2, B(s_1, t) = B(s_2, t)$$

$$z_{nmt}^{s_1} = z_{nmt}^{s_2}, \forall s_1, s_2, n, m, t,$$

$$s_1 \neq s_2, B(s_1, t) = B(s_2, t)$$

$$I_{nt}^{s_1} = I_{nt}^{s_2}, \forall s_1, s_2, n, t,$$

$$s_1 \neq s_2, B(s_1, t) = B(s_2, t)$$

$$x_{nmt}^{s} \geq 0, \forall s, n, m, t$$

$$z_{nmt}^{s} \geq 0, \forall s, n, m, t$$

$$I_{nt}^{s} \geq 0, \forall s, n, t$$

5. NUMERICAL EXPERIMENTS

5.1. Objective and Method of Numerical Experiments

Numerical experiments were performed based on the following perspectives.

1) The benefits of the stochastic programming model are assessed by comparison with the deterministic programming model (see Section 5.2).

2) Next, the behavior of the solution method using parameter ρ, as mentioned in Section 4.3, is analyzed. After that, the performance of the proposed solution method is assessed by comparing it with a case in which a direct branch-and-bound method-based solution method is used for the formulation of the stochastic programming model (a mixed-integer programming problem) (see Section 5.3).

The method of simulation was as follows. The numbers of item types N, machines M, and times T were set as (N, M, T) = (3, 2, 10), (3, 2, 15), (4, 2, 10), (4, 2, 15), (4, 3, 10), (4, 3, 15), (5, 3, 10), (5, 3, 15).

(These problems were then numbered from 1 to 8 so that problems could later be reference by number). Instances were then generated for each of these by random number. However, the changing demand scenarios were applied according to **Table 1**. In the table, up to eight of the T number of times have been divided into four equal sets, which are named P1, P2, P3, and P4 (for example, with T = 10, P1 represents t = 1, 2). The values in the table indicate the ratios of how the demand changes in each scenario, compared with the base amount of demand (the case where s = 1) when the changing demand scenarios were generated.

We consider cases with two, four, and eight scenarios. When S = 2, scenarios 1 and 2 each arise with a probability of $^{1}/_{2}$; when S = 4, scenarios 1, 2, 3 and 4 each arise with a probability of $^{1}/_{4}$. The case when S = 8 is treated similarly.

Table 1. Demand patern in each secenario.

S	P1	P2	P3	P4
1	0	0	0	0
2	0	0	0	+20%
3	0	0	+20%	0
4	0	0	+20%	+20%
5	0	+20%	0	0
6	0	+20%	0	+20%
7	0	+20%	+20%	0
8	0	+20%	+20%	+20%

The package used to apply the scenario aggregation method was OPL Studio 5.2. For optimization of the scenario sub-problem in Step 3, which is a quadratic programming problem having a 0 - 1 condition, and the mixed-integer programming problem used for comparison, the mixed integer optimizer in CPLEX (branchand-bound method-based solution) was used. The computer used for experiments was a DELL Precision 490 (Xeon 5060 3.20 GHz, memory 2 GB).

5.2. Value of Stochastic Solution

As a criterion of assessment, we used the value of stochastic solution (VSS) represented by the equation below. If we take the optimal objective function value of the stochastic programming model to be z_p and the optimal objective function value of the deterministic programming model to be z_d, then VSS is defined as follows.

$$VSS = \frac{z_d - z_p}{z_d} \times 100(\%)$$

The demand at each time in the deterministic programming model was given by the mean of the demand across all scenarios in the stochastic programming model, and we assume that this demand will arise with probability 1 under that scenario only. Specifically, this was solved as one instance of the deterministic programming model, and the values of the decision variables were decided.

Next, we looked at the situation depicted in the formulation below that considers the penalty of running out of inventory, and computed the costs in each scenario when the values of the (only set of) decision variables x_{nmt}, y_{nmt}, z_{nmt} which could solve the current deterministic programming model were used (the number of results obtained corresponds with the number of scenarios). We then computed the mean of these results based on the probabilities of the scenarios arising and used this as the objective function of the deterministic programming model. D^s_{nt} is the amount of demand which could not be satisfied by each scenario, and H_n is the penalty per unit of demand not satisfied (which corresponds to twice the inventory storage cost h_n).

$$\min \sum_s P^s \left(\sum_{n,m,t} c_{nm} z_{nmt} + \sum_{n,t} h_n I^s_{nt} + \sum_{n,t} H_n D^s_{nt} \right) \quad (27)$$

$$\text{s.t. } I^s_{nt} = I^s_{n(t-1)} + \sum_m x_{nmt} - d^s_{nt} + D^s_{nt}, \forall s,n,t \quad (28)$$

$$I^s_{n0} = 0, \forall s,n \quad (29)$$

$$I^s_{nt} \geq 0, \forall s,n,t \quad (30)$$

$$D^s_{nt} \geq 0, \forall s,n,t \quad (31)$$

Factors which do not appear in the original problem are considered here because, in contrast to conventional deterministic and stochastic programming models that do not allow for inventory to run out, when the values of variables are decided in a deterministic programming model that incorporates stochastic variation in the demand, it is possible for inventory to run out. As stated above, the penalty that arises when inventory runs out in such cases is assumed, in this numerical example, to be twice the inventory storage cost.

Table 2 lists the value of solutions from the stochastic programming model. Since it is necessary, in this case, to accurately assess the model, the formulation by the stochastic programming model (a mixed-integer program) was solved in its original format by the mixed integer optimizer in CPLEX, and only the data from which an optimal solution (within 3600 seconds) was obtained was considered. That is to say, the solution found at this time was not from the aforementioned scenario aggregation method. **Table 2** shows the results for Problem 1 and Problem 5, for which optimal solutions could be obtained for all scenarios (as shown in the table).

From **Table 2**, it is apparent that when the number of scenarios increases, the VSS value increases accordingly. This is thought to be due to the fact that when the number of scenarios increases and the element of uncertainty grows, the benefit of the stochastic programming model also increases.

Table 2. Value of solutions from the stochastic model.

	Prob. 1			Prob. 5		
S	stochastic	deterministic	VSS	stochastic	deterministic	VSS
	VOF	VOF	(%)	VOF	VOF	(%)
2	1865.0	1915.5	2.6	1279.5	1330.5	3.8
4	1887.5	2071.5	8.9	1286.0	1456.3	11.7
8	1897.5	2153.5	11.9	1293.5	1564.9	17.3

VOF: value of objective function.

In this section, we have shown only the case solvable as an integer programming problem that is equivalent to the stochastic programming model. However, in general, the scale of the problem grows as the number of scenarios is increased, making it more difficult to obtain a good solution (the optimal solution or one close to the optimal solution). As such, the importance of a procedure for finding the solution for a stochastic programming model directly and in a short period of time is heightened.

5.3. Assessment of Performance of the Proposed Solution Method

The optimal objective function values and computation times for the solution method proposed in the present study and the method of using the mixed integer optimizer in CPLEX directly to determine the formulation of the stochastic programming model (this will be referred to below simply as mixed-integer programming) were compared for the problem groups described in Section 5.1 for two, four, and eight scenarios.

The computation time for the solution method of the present study and mixed-integer programming were both limited to less than 3600 seconds; if the optimal value was not obtained by mixed-integer programming, a tentative objective function value was computed when the computation time reached the limit. The results are shown in Tables 3-5.

With proposed solution method, it was necessary to set the value of the parameter ρ that appears in Step 3 of Section 4.3. It was predicted that the behavior of the optimization would depend on this parameter value. Therefore, all of the results are shown for the values $\rho = 0.2, 0.4, 0.6, 0.8$, and 1.0. The optimization problems that appear in each step of the proposed solution method, as described in Section 4.3, were solved by using CPLEX as stated above.

Two points are of interest here:

- The relationship between the parameter set and the behavior of the solution method
- The performance of the proposed solution method in contrast to the case where the mixed-integer programming problem was solved directly using the branch-and-bound method-based solution

Table 3. Results from the stochastic model (# of scenario = 2).

	ρ										MIP	
	0.2		0.4		0.6		0.8		1.0			
prob.	sol.	time	sol.	time	sol	time	sol.	time	sol.	time	sol.	time
1	1865.0	11	1865.0	11	1865.0	11	1865.0	11	1865.0	11	1865.0	5
2	2853.0	18	2853.0	18	2853.0	18	2853.0	18	2853.0	18	2853.0	8
3	3149.5	18	3149.5	18	3149.5	18	3149.5	18	3149.5	18	3149.5	8
4	3901.8	56	3901.8	56	3901.8	56	3901.8	56	3901.8	56	3891.2	55
5	1279.5	13	1279.5	13	1279.5	13	1279.5	13	1279.5	13	1279.5	7
6	2768.0	29	2758.0	29	2768.0	29	2768.0	29	2768.0	29	2768.0	21
7	2964.5	44	2964.5	44	2964.5	44	2964.5	44	2964.5	44	2964.5	35
8	3023.3	62	3023.3	55	3023.3	62	3023.3	62	3023.3	55	3023.3	53

Table 4. Results from the stochastic model (# of scenario = 4).

	ρ										MIP	
	0.2		0.4		0.6		0.8		1.0			
prob.	sol.	time	sol.	time	sol	time	sol.	time	sol.	time	sol.	time
1	1887.5	20	1887.5	20	1887.5	20	1887.5	20	1887.5	20	1887.5	6
2	2955.0	64	2955.0	64	2955.0	64	2955.0	86	2955.0	108	2892.5	20
3	3167.0	45	3167.0	35	3167.0	35	3167.0	35	3167.0	35	3167.0	13
4	3951.4	114	3951.4	114	3951.4	114	3951.4	114	3951.4	114	3940.8	352
5	1286.0	24	1286.0	24	1286.0	24	1286.0	24	1286.0	24	1286.0	9
6	2849.0	72	3063.0	58	3063.0	58	3063.0	59	3063.0	59	2835.7	89
7	3023.0	82	3023.0	82	3023.0	82	3023.0	82	3023.0	82	3017.0	213
8	3120.3	195	3120.3	110	3120.3	110	3120.3	110	3264.3	98	3120.3	322

Table 5. Results from the stochastic model (# of scenario = 8).

	ρ											
	0.2		0.4		0.6		0.8		1.0		MIP	
prob.	sol.	time	sol.	time	sol	time	sol.	time	sol.	time	sol.	time
1	1897.5	41	1897.5	41	1897.5	41	1897.5	41	1897.5	41	1897.5	7
2	3010.0	118	3010.0	118	3010.0	118	3010.0	167	3010.0	207	2916.5	857
3	3185.0	59	3185.0	59	3185.0	59	3185.0	59	3185.0	59	3185.0	146
4	3999.1	217	3999.1	217	3999.1	217	3999.1	217	3999.1	217	4071.6	3600
5	1293.5	48	1293.5	48	1293.5	48	1293.5	48	1293.5	48	1293.5	30
6	2903.0	135	3010.0	101	3010.0	101	3010.0	101	3010.0	101	2903.0	3600
7	3054.5	156	3054.5	156	3054.5	156	3054.5	156	3054.5	156	3048.5	3600
8	3199.5	250	3199.5	250	3440.8	249	3440.8	249	3349.8	218	3198.8	3600

The changes in behavior that result from changing the parameter ρ can be observed by comparing within a given row of a particular table. However, no specific pattern of behavior can necessarily be observed in the computation times and the optimal objective function values, and even when the parameter is changed in addressing the same problem, one cannot say that the effect is large enough to bring about any significant differences.

Comparing the solution method proposed in the present study and mixed-integer programming using Tables 3-5 reveals that although the proposed solution method does not always compute the same result as mixed-integer programming, in many cases either the same or an extremely close value is obtained.

There is a striking increase in the computation time of mixed-integer programming as the number of scenarios increases. In contrast, the computation time of the solution method we propose is relatively short, and although it falls a little behind the results of the latter in terms of how good the solution is, in cases where there are many scenarios, it is overwhelmingly superior in computation time. In particular, when the number of scenarios is high ($S = 8$), the proposed solution method obtains the feasible solution in a short period of time, in contrast to the mixed-integer programming computation, which frequently reaches the upper limit of time and has to stop.

This difference in computation time trends is thought to be due to the fact that, with the solution method we propose, although the computation time grows as the number of scenarios increases, the fact that the problem is broken down into sub-problems for each scenario which are then solved means that the effects of increasing the number of scenarios can be limited, compared to mixed-integer programming.

6. CONCLUSION AND FUTURE CHALLENGES

In the present study, we considered the formulation of the lot-scheduling problem on parallel machines using a stochastic programming model and demonstrated the benefit of such a model over a deterministic programming model.

We developed an approximate solution method which applied the scenario aggregation method and demonstrated that even when the number of scenarios increases, thus making the problem large in scale, it is possible to compute an accurate solution that is of practical application in a short period of time.

REFERENCES

1. J. R. Birge, "Stochastic Programming Computation and Applications," INFORMS Journal on Computing, Vol. 9, No. 2, 1997, pp. 111-133.

2. J. R. Birge and F. Louveaux, "Introduction to Stochastic Programming," Springer-Verlag, Berlin, 1997.

3. T. Shiina, "Stochastic Programming (in Japanese)," In: M. Kubo, A. Tamura and T. Matsui, Eds., Ôyô Sûri-Keikaku Handbook, Asakura Syoten, Tokyo, 2002, pp. 710-769.

4. R. T. Rockafellar and R. J.-B. Wets, "Scenarios and Policy Aggregation in Optimization under Uncertainty," Mathematics of Operations Research, Vol. 16, No. 1, 1991, pp. 119-147.

5. A. Løkketangen and D. L. Woodruff, "Progressive Hedging and Tabu Search Applied to Mixed Integer (0,1) Multistage Stochastic Programming," Journal of Heuristics, Vol. 2, 1996, pp. 111-128.

6. H. Meyr, "Simultaneous Lotsizing and Scheduling on Parallel Machines," European Journal of Operational Research, Vol. 139, No. 2, 2002, pp. 277-292.

7. H. Arai, S. Morito and J. Imaizumi, "A Column Generation Approachfor Discrete Lotsizing and Scheduling Problem on Identical Parallel Machines," Journal of Japan Industrial Management Association, Vol. 55, 2004, pp. 69-76.

5

CHAPTER

Solution of Stochastic Cubic and Quintic Nonlinear Diffusion Equation Using WHEP, Pickard and HPM Methods

Magdy A. El-Tawil[1], Aisha F.Fareed[2]

[1]*Cairo university, Faculty of Engineering, Engineering, Mathematics Department, Giza, Egypt*

[2]*Benha university, Benha Higher Institute of Technology, Basic Science Department, Benha, Egypt*

ABSTRACT

In this paper, the cubic and quintic diffusion equation under stochastic non homogeneity is solved using WienerHermite expansion and perturbation (WHEP) technique, Homotopy perturbation method (HPM) and Pickard approximation technique. The analytic solution of the linear case is obtained using Eigenfunction expansion .The Picard approximation method is used to introduce the first and second order approximate solution for the non linear case. The WHEP technique is also used to obtain approximate solution under different orders and different corrections. The Homotopy perturbation method (HPM) is also used to obtain some approximation orders for mean and variance. Using mathematica-5, the methods of solution are illustrated through figures, comparisons among different methods and some parametric studies.

Keywords: Stochastic diffusion equations, WHEP technique, HPM method

1. INTRODUCTION

The study of random solutions of partial differential equations was initiated by Kampe de Feriet in 1955 [1]. In his valuable survey on the theory of random equations, Bharucha-Reid showed how a stochastic heat equation of Cauchy type can be solved using the stochastic integrals theory[2]. In 1973, Lo Dato V. [3] considered the stochastic velocity field and the Navier-Stokes equation and discussed the mathematical problems associated with it. Becus A. Georges [4] introduced a general solution for the heat conduction problem with a random source term and random initial and boundary conditions. Many authors investigated the stochastic diffusion equation under different views, see [5-11].

El-Tawil M. used the Wiener-Hermite expansion together with perturbation theory (WHEP technique) to solve a perturbed nonlinear stochastic diffusion equation [12]. The technique has been then developed to be applied on non-perturbed differential equations using the homotopy perturbation method and is called Homotopy WHEP [13,14]. El-Tawil M. and Noha A. El-Molla.[15] solved the quadratic and cubic non-linear stochastic diffusion equation using Pickard approximation and homotopy WHEP technique [16].

The diffusion equation with cubic and quintic nonlinear losses and stochastic non homogeneity are solved using different techniques, mainly the Pickard approximation, the WHEP technique and HPM. The main goal of the paper is to compare among these different techniques. Some statistical moments are obtained, mainly the ensemble average, covariance and variance of the solution processes. In Sections 2.1 and 3.1, the Pickard approximation technique is used in solving the cubic and quintic nonlinear diffusion problems respectively. WHEP technique is processed in Sections 2.2 and 3.2, while HPM is used in Sections 2. 3 and 3. 3. Some comparisons are illustrated in different sections.

2. THE CUBIC NONLINEAR STOCHASTIC DIFFUSION EQUATION

Let us consider the following stochastic nonlinear-diffusion equation with cubic nonlinear losses, $\varepsilon \cdot u^3$:

$$\frac{\partial u(x,t)}{\partial t} = \frac{\partial^2 u}{\partial x^2} - \varepsilon \cdot u^2 + \sigma \cdot n(x);$$
$$(x,t) \in (0,L) \times (0,\infty),\ u(0,t) = 0,\ u(L,t) = 0$$

$$\text{and} \quad u(x,0) = \varphi(x). \tag{1}$$

Where ε is a deterministic scale for the nonlinear term. The in homogeneity term $\sigma \cdot n(x;\omega)$ is space white noise scaled by σ.

Three methods are used in the next subsections, mainly the Pickard approximations, the WHEP technique and Homotopy perturbation method.

2.1. Using Pickard Approximation

In this technique, the linear part of the differential operator is kept in the left hand side of the equation whereas the rest of the nonlinear terms are moved to the right part. The successive Pickard approximations are processed according to letting the L.H.S. as the $n+1$ approximation for the solution process depending on the nth approximation in the R.H.S, $n \geq 0$ Following this routine and applying it on to (1), we get the following iterative equations:

$$\frac{\partial u_0}{\partial t}-\frac{\partial^2 u_0}{\partial x^2}-\sigma n(x;\omega)=0$$
$$u_0(0,x)=\varphi(x),\ u_0(t,0)=u_0(t,L)=0 \tag{2}$$

$$\frac{\partial u_{n+1}}{\partial t}-\frac{\partial^2 u_{n+1}}{\partial x^2}-\sigma n(x;\omega)=-\varepsilon u_n,\ n\geq 0,$$
$$u_{n+1}(0,x)=\varphi(x),\ u_{n+1}(t,0)=u_{n+1}(t,L)=0 \tag{3}$$

Using Eigen function expansion, the following general solutions are got

$$u_0(t,x)=\sum_{n=0}^{\infty}T_{n0}\mathrm{e}^{-\left(\frac{n\pi}{L}\right)^2}\sin\frac{n\pi}{L}x+\sum_{n=0}^{\infty}I_{n0}(t)\sin\frac{n\pi}{L}x \tag{4}$$

where

$$F_{n_0}(t)=\frac{2\sigma}{L}\int_0^L n(x)\sin\frac{n\pi}{L}x\,\mathrm{d}x, \tag{5}$$

$$I_{n_0}(t)=\int_0^t \mathrm{e}^{-\left(\frac{n\pi}{L}\right)^2(t-\tau)}F_{n_0}(\tau)\,\mathrm{d}\tau \tag{6}$$

Also,

$$u_{n+1}(t,x)=\sum_{m=0}^{\infty}T_{n_{n+1}}\mathrm{e}^{-\left(\frac{m\pi}{L}\right)^2 t}\sin\frac{m\pi}{L}x+\sum_{m=0}^{\infty}I_{n_{n+1}}(t)\sin\frac{m\pi}{L}x \tag{7}$$

where

$$F_{n_{n+1}}(t)=\frac{2\sigma}{L}\int_0^L n(x)\sin\frac{m\pi}{L}x\,dx$$
$$-\frac{2\varepsilon}{L}\int_0^L u_n^3(t,x)\sin\frac{m\pi}{L}x\,dx \tag{8}$$

$$I_{n_{n+1}}(t)=\int_0^t e^{-(\frac{m\pi}{L})^2(t-\tau)}F_{n_{n+1}}(\tau)\,d\tau \tag{9}$$

$$T_{n_{n+1}}=\frac{2}{L}\int_0^L \varphi(x)\sin\frac{m\pi}{L}x\,dx. \tag{10}$$

If the convergence of the process is insured, one can obtain the solution as

$$u(t,x)=\lim_{n\to\infty}u_n(t,x) \tag{11}$$

One can notice that all order of approximations are stochastic processes. The ensemble average of the zero order approximation is obtained as

$$\mu_{u_0}(t,x)=\sum_{n=0}^{\infty}T_{n_0}e^{-\left(\frac{m\pi}{L}\right)^2 t}\sin\frac{m\pi}{L}x. \tag{12}$$

The covariance of u_0 is given by

$$Cov\left(u_0\left(t,x_1\right),u_0\left(t,x_2\right)\right)=\frac{4\sigma^2}{L^2}\sum_{n=0}^{\infty}\sum_{m=0}^{\infty}I_{n,m}\left(\int_0^t \mathrm{e}^{-\left(\frac{n\pi}{L}\right)^2(t-\tau)}\mathrm{d}\tau\right)\left(\int_0^t \mathrm{e}^{-\left(\frac{m\pi}{L}\right)^2(t-\tau)}\mathrm{d}\tau\right)\sin\frac{n\pi}{L}x_1\sin\frac{m\pi}{L}x_2 \tag{13}$$

where

$$I_{n,m}=\int_0^L \sin\frac{n\pi}{L}x\sin\frac{m\pi}{L}x\,\mathrm{d}x \tag{14}$$

The variance is

$$\sigma_{u_0}^2\left(t,x\right)=\frac{4\sigma^2}{L^2}\sum_{n=0}^{\infty}\sum_{m=0}^{\infty}I_{n,m}\left(\int_0^t \mathrm{e}^{-\left(\frac{n\pi}{L}\right)^2(t-\tau)}\mathrm{d}\tau\right)\left(\int_0^t \mathrm{e}^{-\left(\frac{m\pi}{L}\right)^2(t-\tau)}\mathrm{d}\tau\right)\sin\frac{n\pi}{L}x\sin\frac{m\pi}{L}x. \tag{15}$$

The following results for the **first order approximation** are obtained:

$$u_1(t,x)=\sum_{n=0}^{\infty}T_{n_0}e^{-\left(\frac{n\pi}{L}\right)^2 t}\sin\frac{n\pi}{L}x$$
$$+\sum_{n=0}^{\infty}\left(\int_0^t e^{-\left(\frac{n\pi}{L}\right)^2(t-\tau)}F_{n_1}(\tau)\mathrm{d}\tau\right)\sin\frac{n\pi}{L}x \quad (16)$$

$$F_{n_1}(t)=\frac{2\sigma}{L}\int_0^L n(x)\sin\frac{n\pi}{L}x-\frac{2\varepsilon}{L}\int_0^L\left[u_0(t,x)\right]^3\sin\frac{n\pi}{L}x\,\mathrm{d}x \quad (17)$$

where

$$\left[u_0(t,x)\right]^3=\mu_{u_0}^3(t,x)+3\mu_{u_0}^2(t,x)\sum_{n=0}^{\infty}I_{n_0}(t)\sin\frac{n\pi}{L}x$$
$$+3\mu_{u_0}(t,x)\sum_{n=0}^{\infty}\sum_{m=0}^{\infty}I_{n_0}(t)I_{m_0}(t)\sin\frac{n\pi}{L}x$$
$$\sin\frac{m\pi}{L}x$$
$$+\sum_{n=0}^{\infty}\sum_{m=0}^{\infty}\sum_{l=0}^{\infty}I_{n_0}(t)I_{m_0}(t)I_{l_0}(t)\sin\frac{n\pi}{L}x$$
$$\sin\frac{m\pi}{L}x\sin\frac{l\pi}{L}x \quad (18)$$

$$\mu_{u_1}(t,x)=\sum_{n=0}^{\infty}T_{n_0}e^{-\left(\frac{n\pi}{L}\right)^2 t}\sin\frac{n\pi}{L}x$$
$$+\sum_{n=0}^{\infty}\left(\int_0^t e^{-\left(\frac{n\pi}{L}\right)^2(t-\tau)}\mu_{F_{n_1}}(\tau)\mathrm{d}\tau\right)\sin\frac{n\pi}{L}x \quad (19)$$

where

$$\mu_{F_{n_1}}(t)=\frac{-2\varepsilon}{L}\int_0^L\left(\mu_{u_0}^3(t,x)+3\mu_{u_0}(t,x)\sigma_{u_0}^2(t,x)\right)$$
$$\sin\frac{n\pi}{L}x\,\mathrm{d}x \tag{20}$$

$$\begin{aligned}
&Cov\left(u_1(t,x_1),u_1(t,x_2)\right)\\
&=\sum_{n=0}^{\infty}\sum_{m=0}^{\infty}EI_{n_1}(t)I_{m_1}(t)\sin\frac{n\pi}{L}x_1\sin\frac{m\pi}{L}x_2\\
&\quad-\sum_{n=0}^{\infty}\left(\int_0^t e^{-\left(\frac{n\pi}{L}\right)^2(t-\tau)}EF_{n_1}(\tau)\,\mathrm{d}\tau\right)\sin\frac{n\pi}{L}x_2\\
&\quad\cdot\sum_{m=0}^{\infty}EI_{m_1}(t)\sin\frac{m\pi}{L}x_1\\
&\quad-\sum_{m=0}^{\infty}\left(\int_0^t e^{-\left(\frac{m\pi}{L}\right)^2(t-\tau)}EF_{m_1}(\tau)\,\mathrm{d}\tau\right)\sin\frac{m\pi}{L}x_1\\
&\quad\cdot\sum_{n=0}^{\infty}EI_{n_1}(t)\sin\frac{n\pi}{L}x_2\\
&\quad+\sum_{n=0}^{\infty}EI_{n_1}(t)\sin\frac{n\pi}{L}x_1\cdot\sum_{m=0}^{\infty}EI_{m_1}(t)\sin\frac{m\pi}{L}x_2
\end{aligned} \tag{21}$$

where E denotes the ensemble average operator and

$$EI_{n_1}(t)=\int_0^t e^{-\left(\frac{n\pi}{L}\right)^2(t-\tau)}EF_{n_1}(\tau)\,\mathrm{d}\tau \tag{22}$$

$$EI_{n_1}(t)I_{m_1}(t)$$
$$=\int_0^t\int_0^t e^{-\left(\frac{n\pi}{L}\right)^2(t-\tau_1)} e^{-\left(\frac{m\pi}{L}\right)^2(t-\tau_2)} EF_{n_1}(\tau_1)F_{m_1}(\tau_2)\,d\tau_1 d\tau_2 \tag{23}$$

In which

$$EF_{n_1}(\tau_1)F_{m_1}(\tau_2)$$
$$=\frac{4\sigma^2}{L^2}\int_0^L \sin\frac{n\pi}{L}x\sin\frac{m\pi}{L}x\,dx$$
$$-\frac{4\varepsilon\sigma}{L^2}\int_0^L\int_0^L \sin\frac{n\pi}{L}x_3\sin\frac{m\pi}{L}x_4\; En(x_3)u_0^3(\tau_2,x_4)\,dx_3 dx_4$$
$$-\frac{4\varepsilon\sigma}{L^2}\int_0^L\int_0^L \sin\frac{n\pi}{L}x_3\sin\frac{m\pi}{L}x_4\; En(x_4)u_0^3(\tau_1,x_3)\,dx_3 dx_4$$
$$+\frac{4\varepsilon^2}{L^2}\int_0^L\int_0^L \sin\frac{n\pi}{L}x_3\sin\frac{m\pi}{L}x_4\; Eu_0^3(\tau_1,x_3)u_0^3(\tau_2,x_4)\,dx_3 dx_4 \tag{24}$$

where

$$En(x)u_0^3(t,y)$$
$$=3\left(\sum_{n=0}^{\infty}T_{n_0}e^{-\left(\frac{n\pi}{L}\right)^2 t}\sin\frac{n\pi}{L}y\right)^2\sum_{n=0}^{\infty}\sin\frac{n\pi}{L}y\,En(x)I_{n_0}(t)$$
$$+\sum_{n=0}^{\infty}\sum_{m=0}^{\infty}\sum_{l=0}^{\infty}En(x)I_{n_0}(t)I_{m_0}(t)I_{l_0}(t)\sin\frac{n\pi}{L}x$$
$$\sin\frac{m\pi}{L}x\sin\frac{l\pi}{L}x \tag{25}$$

In which

$$En(x)I_{n_0}(t)=\frac{2\sigma}{L}\sin\frac{n\pi}{L}x\int_0^t e^{-\left(\frac{n\pi}{L}\right)^2(t-\tau)}d\tau \tag{26}$$

$$\begin{aligned}&En(x)I_{n_0}(t)I_{m_0}(t)I_{l_0}(t)\\&=\int_0^t\int_0^t\int_0^t e^{-\left(\frac{n\pi}{L}\right)^2(t-\tau_1)}e^{-\left(\frac{m\pi}{L}\right)^2(t-\tau_2)}e^{-\left(\frac{l\pi}{L}\right)^2(t-\tau_3)}\\&\quad En(x)F_{n_0}(\tau_1)F_{m_0}(\tau_2)F_{l_0}(\tau_3)d\tau_1 d\tau_2 d\tau_3\end{aligned} \tag{27}$$

And

$$\begin{aligned}&Eu_0^3(\tau_1,x_3)u_0^3(\tau_2,x_4)\\&=\phi^3[\tau_1,x_3]\phi^3[\tau_2,x_4]+3\phi^3[\tau_1,x_3]\phi[\tau_2,x_4]\\&\sum_{n=0}^{\infty}\sum_{m=0}^{\infty}EI_{n_0}(\tau_2)I_{m_0}(\tau_2)\sin\frac{n\pi}{l}x_4\sin\frac{m\pi}{l}x_4\\&\quad+3\phi^3[\tau_2,x_4]\phi[\tau_1,x_3]\sum_{n=0}^{\infty}\sum_{m=0}^{\infty}EI_{n_0}(\tau_1)I_{m_0}(\tau_1)\\&\quad\sin\frac{n\pi}{l}x_3\sin\frac{m\pi}{l}x_3\\&\quad+9\phi^2[\tau_1,x_3]\phi^2[\tau_2,x_4]\sum_{n=0}^{\infty}\sum_{m=0}^{\infty}EI_{n_0}(\tau_1)I_{m_0}(\tau_2)\\&\quad\sin\frac{n\pi}{l}x_3\sin\frac{m\pi}{l}x_4\\&\quad+3\phi^2[\tau_1,x_3]\sum_{n=0}^{\infty}\sum_{m=0}^{\infty}\sum_{l=0}^{\infty}\sum_{k=0}^{\infty}EI_{n_0}(\tau_1)I_{m_0}(\tau_2)\,I_{l_0}(\tau_2)\end{aligned}$$

$$I_{k_0}(\tau_2)\sin\frac{n\pi}{l}x_3\sin\frac{m\pi}{l}x_4\sin\frac{l\pi}{l}x_4\sin\frac{k\pi}{l}x_4$$

$$+3\phi^2[\tau_2,x_4]\sum_{n=0}^{\infty}\sum_{m=0}^{\infty}\sum_{l=0}^{\infty}\sum_{k=0}^{\infty}EI_{n_0}(\tau_1)I_{m_0}(\tau_1)I_{l_0}(\tau_1)$$

$$I_{k_0}(\tau_2)\sin\frac{n\pi}{l}x_3\sin\frac{m\pi}{l}x_3\sin\frac{l\pi}{l}x_3\sin\frac{k\pi}{l}x_4$$

$$+9\phi[\tau_1,x_3]\phi[\tau_2,x_4]\sum_{n=0}^{\infty}\sum_{m=0}^{\infty}\sum_{l=0}^{\infty}\sum_{k=0}^{\infty}EI_{n_0}(\tau_1)$$

$$I_{m_0}(\tau_1)I_{l_0}(\tau_2)I_{k_0}(\tau_2)\sin\frac{n\pi}{l}x_3\sin\frac{m\pi}{l}x_3$$

$$\sin\frac{l\pi}{l}x_4\sin\frac{k\pi}{l}x_4$$

$$+\sum_{n=0}^{\infty}\sum_{m=0}^{\infty}\sum_{l=0}^{\infty}\sum_{k=0}^{\infty}\sum_{v=0}^{\infty}\sum_{w=0}^{\infty}EI_{n_0}(\tau_1)I_{m_0}(\tau_1)I_{l_0}(\tau_1)$$

$$I_{k_0}(\tau_2)I_{v_0}(\tau_2)I_{w_0}(\tau_2)\sin\frac{n\pi}{l}x_3\sin\frac{m\pi}{l}x_3\sin\frac{l\pi}{l}x_3$$

$$\sin\frac{k\pi}{l}x_4\sin\frac{v\pi}{l}x_4\sin\frac{w\pi}{l}x_4 \tag{28}$$

In which

$$\phi(t,x)=\sum_{n=0}^{\infty}T_{n_0}e^{-\left(\frac{n\pi}{L}\right)^2 t}\sin\frac{n\pi}{L}x, \tag{29}$$

$$EI_{n_0}(\tau_1)I_{m_0}(\tau_1)$$
$$=\frac{4\sigma^2}{L^2}I_{n,m}\int_0^{\tau_1}\int_0^{\tau_1}e^{-\left(\frac{n\pi}{L}\right)^2(\tau_1-\tau_3)}e^{-\left(\frac{m\pi}{L}\right)^2(\tau_1-\tau_4)}d\tau_3 d\tau_4 \tag{30}$$

$$EI_{n_0}(\tau_2)I_{m_0}(\tau_2)$$
$$=\frac{4\sigma^2}{L^2}I_{n,m}\int_0^{\tau_2}\int_0^{\tau_2}e^{-\left(\frac{n\pi}{L}\right)^2(\tau_2-\tau_3)}e^{-\left(\frac{m\pi}{L}\right)^2(\tau_2-\tau_4)}d\tau_3 d\tau_4 \tag{31}$$

$$EI_{n_0}(\tau_1)I_{m_0}(\tau_2) = \frac{4\sigma^2}{L^2} I_{n,m} \int_0^{\tau_2}\int_0^{\tau_1} e^{-\left(\frac{n\pi}{L}\right)^2(\tau_1-\tau_3)} e^{-\left(\frac{m\pi}{L}\right)^2(\tau_2-\tau_4)} d\tau_3 d\tau_4 \tag{32}$$

$$\begin{aligned} &EI_{n_0}(\tau_1)I_{m_0}(\tau_1)I_{l_0}(\tau_1)I_{k_0}(\tau_2) \\ &= \int_0^{\tau_2}\int_0^{\tau_1}\int_0^{\tau_1}\int_0^{\tau_1} e^{-\left(\frac{n\pi}{L}\right)^2(\tau_1-\tau_3)} e^{-\left(\frac{m\pi}{L}\right)^2(\tau_1-\tau_4)} e^{-\left(\frac{l\pi}{L}\right)^2(\tau_1-\tau_5)} e^{-\left(\frac{k\pi}{L}\right)^2(\tau_2-\tau_6)} \\ &EF_{n_0}(\tau_3)F_{m_0}(\tau_4)F_{l_0}(\tau_5)F_{k_0}(\tau_6)d\tau_3 d\tau_4 d\tau_5 d\tau_6, \end{aligned} \tag{33}$$

$$\begin{aligned} &EI_{n_0}(\tau_1)I_{m_0}(\tau_2)I_{l_0}(\tau_2)I_{k_0}(\tau_2) \\ &= \int_0^{\tau_2}\int_0^{\tau_2}\int_0^{\tau_2}\int_0^{\tau_1} e^{-\left(\frac{n\pi}{L}\right)^2(\tau_1-\tau_3)} e^{-\left(\frac{m\pi}{L}\right)^2(\tau_1-\tau_4)} e^{-\left(\frac{l\pi}{L}\right)^2(\tau_2-\tau_5)} e^{-\left(\frac{k\pi}{L}\right)^2(\tau_2-\tau_6)} \\ &EF_{n_0}(\tau_3)F_{m_0}(\tau_4)F_{l_0}(\tau_5)F_{k_0}(\tau_6)d\tau_3 d\tau_4 d\tau_5 d\tau_6, \end{aligned} \tag{34}$$

$$\begin{aligned} &EI_{n_0}(\tau_1)I_{m_0}(\tau_1)I_{l_0}(\tau_2)I_{k_0}(\tau_2) \\ &= \int_0^{\tau_1}\int_0^{\tau_1}\int_0^{\tau_2}\int_0^{\tau_2} e^{-\left(\frac{n\pi}{L}\right)^2(\tau_1-\tau_3)} e^{-\left(\frac{m\pi}{L}\right)^2(\tau_1-\tau_4)} e^{-\left(\frac{l\pi}{L}\right)^2(\tau_2-\tau_5)} e^{-\left(\frac{k\pi}{L}\right)^2(\tau_2-\tau_6)} \\ &EF_{n_0}(\tau_3)F_{m_0}(\tau_4)F_{l_0}(\tau_5)F_{k_0}(\tau_6)d\tau_3 d\tau_4 d\tau_5 d\tau_6, \end{aligned} \tag{35}$$

$$EI_{n_0}(\tau_1)I_{m_0}(\tau_1)I_{l_0}(\tau_1)I_{k_0}(\tau_2)I_{v_0}(\tau_2)I_{w_0}(\tau_2)$$

$$= \int_0^{\tau_2}\int_0^{\tau_2}\int_0^{\tau_2}\int_0^{\tau_1}\int_0^{\tau_1}\int_0^{\tau_1} e^{-\left(\frac{n\pi}{L}\right)^2(\tau_1-\tau_3)} e^{-\left(\frac{m\pi}{L}\right)^2(\tau_1-\tau_4)} e^{-\left(\frac{l\pi}{L}\right)^2(\tau_1-\tau_5)} e^{-\left(\frac{k\pi}{L}\right)^2(\tau_2-\tau_6)}$$

$$e^{-\left(\frac{v\pi}{L}\right)^2(\tau_2-\tau_7)} e^{-\left(\frac{w\pi}{L}\right)^2(\tau_2-\tau_8)}$$

$$EF_{n_0}(\tau_3)F_{m_0}(\tau_4)F_{l_0}(\tau_5)F_{k_0}(\tau_6)F_{v_0}(\tau_7)$$

$$F_{w_0}(\tau_8)\,d\tau_3 d\tau_4 d\tau_5 d\tau_6 d\tau_7 d\tau_8, \tag{36}$$

where

$$EF_{n_0}(\tau_3)F_{m_0}(\tau_4)F_{l_0}(\tau_5)F_{k_0}(\tau_6)$$

$$= \frac{16\sigma^4}{L^4}\int_0^L\int_0^L \sin\frac{n\pi}{L}x_1 \sin\frac{m\pi}{L}x_1 \sin\frac{l\pi}{L}x_2 \sin\frac{k\pi}{L}x_2 \, dx_1 dx_2$$

$$+ \frac{16\sigma^4}{L^4}\int_0^L\int_0^L \sin\frac{n\pi}{L}x_1 \sin\frac{l\pi}{L}x_1 \sin\frac{m\pi}{L}x_2 \sin\frac{k\pi}{L}x_2 \, dx_1 dx_2$$

$$+ \frac{16\sigma^4}{L^4}\int_0^L\int_0^L \sin\frac{n\pi}{L}x_1 \sin\frac{k\pi}{L}x_1 \sin\frac{m\pi}{L}x_2 \sin\frac{l\pi}{L}x_2 \, dx_1 dx_2. \tag{37}$$

Using mathematica-5, the previous huge computations were performed and the following sample results are obtained:

Figures 1 and 2 illustrates the change of the first order mean and variance under the change of some illustrated parameters.

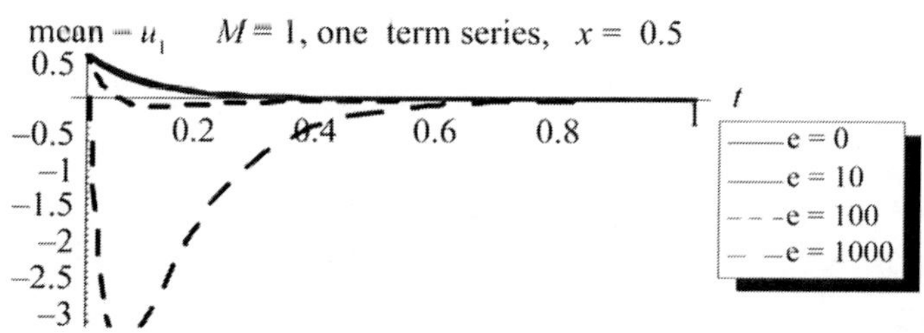

Figure 1. The change of the mean of first order approximation u_1 with time t at x = 0.1 with different ε values, (L = 1, M = 1, σ = 1, Φ(x) = x).

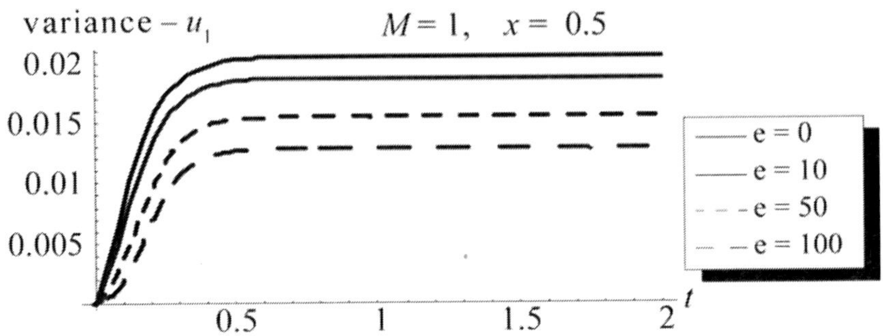

Figure 2. The change of the variance of the first order approximation $u_1$with time t and space variable x at x = 0.5 at different ε values, (L = 1, M = 1, σ = 1, Φ(x) = x).

One can notice that the mean diminishes with time for all ε while the variance decreases with the increase of ε.

Following similar computation procedure, the following results for the second order approximations of the mean is illustrated in figures 3, 4, 5 and 6.

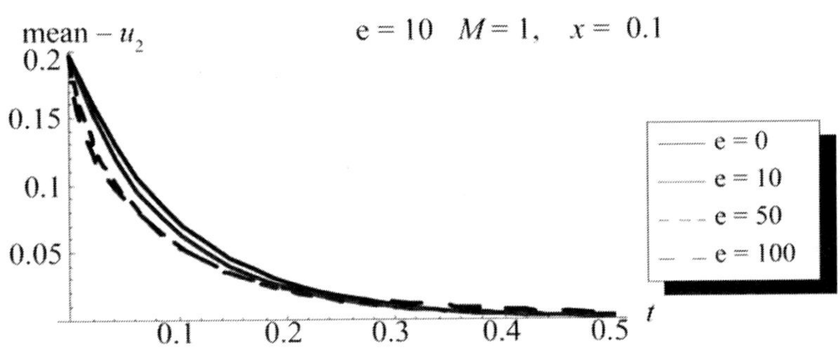

Figure 3. The change of the mean of second order approximation u_2 with time t at x = 0.1 with different ε values, (L = 1, M = 1, σ = 1, Φ(x) = x).

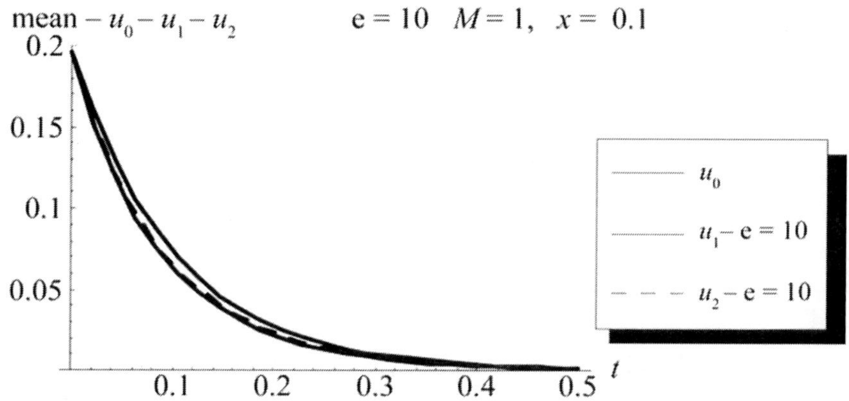

Figure 4. The change of the mean of zero, first and second order approximation (u_0, u_1, u_2) with time t at x = 0.1, ε = 10, (L = 1, M = 1, σ = 1, Φ(x) = x).

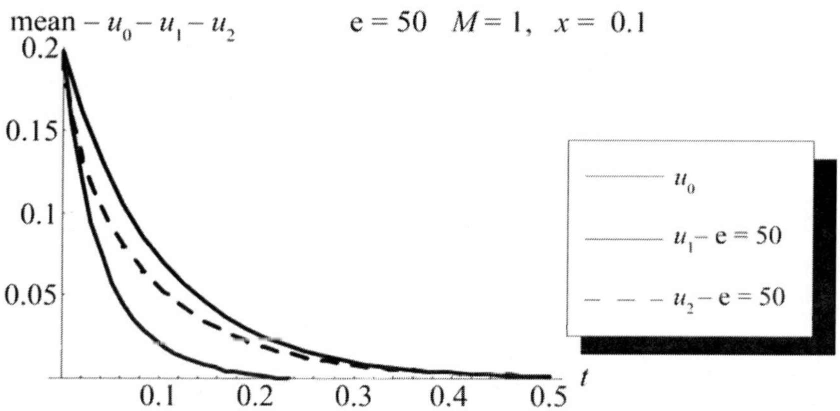

Figure 5. The change of the mean of zero, first and second order approximation (u_0, u_1, u_2) with time t at x = 0.1, ε = 50, (L = 1, M = 1, σ = 1, Φ(x) = x).

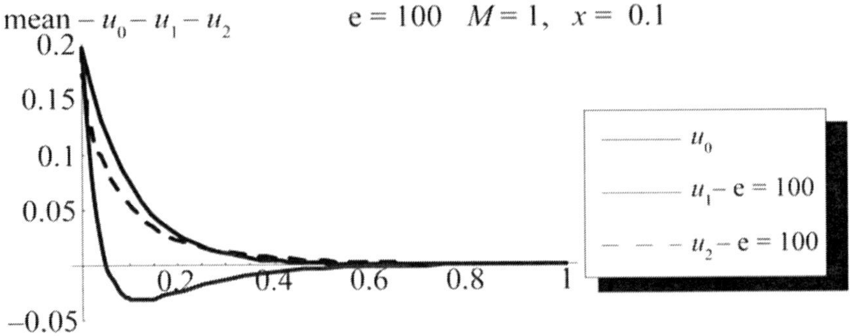

Figure 6. The change of the mean of zero, first and second order approximation (u_0, u_1, u_2) with time t at x = 0.1, ε = 100, (L = 1, M = 1, υ – 1, Φ(x) – x).

2.2. Using WHEP Technique

Since Meecham and his co-workers [25] developed a theory of turbulence involving a truncated Wiener-Hermite expansion (WHE) of the velocity field, many authors studied problems concerning turbulence [26-27]. A lot of general applications in fluid mechanics was also studied in [28]. Scattering problems attracted the WHE applications through many authors [29]. The nonlinear oscillators were considered as an opened area for the applications of WHE as can be found in [30]. There are a lot of applications in boundary value problems [31] and generally in different mathematical studies [32].

The application of the WHE aims at finding a truncated series solution to the solution process of differential equations. The truncated series composes of two major parts; the first is the Gaussian part which consists of the first two terms, while the rest of the series constitute the non-Gaussian part. In nonlinear cases, there exist always difficulties of solving the resultant set of deterministic integro-differential equations got from the applications of a set of comprehensive averages on the stochastic integro-differential equation obtained after the direct application of WHE. Many authors introduced different methods to face these obstacles. Among them, the WHEP technique was introduced in [33] using the perturbation technique to solve perturbed nonlinear problems.

The WHE method utilizes the Wiener-Hermite polynomials which are the elements of a complete set of statistically orthogonal random functions [34]. The WienerHermite polynomial $H^{(i)}(t_1,t_2,\cdots t_i)$ satisfies the following recurrence relation:

$$\begin{aligned} &H^{(i)}(t_1,t_2,\cdots t_i) \\ &= H^{(i-1)}(t_1,t_2,\cdots t_{i-1}).H^{(1)}(t_i) \\ &\quad -\sum_{m=1}^{i-1} H^{(i-2)}(t_1,t_2,\cdots t_{i-2})\cdot\delta(t_{i-m}-t_i),\ i\geq 2 \end{aligned} \tag{38}$$

where

$$\begin{aligned} &H^{(0)} = 1, \\ &H^{(1)}(t) = n(t), \\ &\mathrm{H}^{(2)}(t_1,t_2) = H^{(1)}(t_1)\cdot H^{(1)}(t_2) - \delta(t_1,t_2), \\ &H^{(3)}(t_1,t_2,t_3) = \mathrm{H}^{(2)}(t_1,t_2)\cdot H^{(1)}(t_3) - H^{(1)}(t_1) \\ &\quad \cdot\delta(t_2-t_3) - H^{(1)}(t_2)\cdot\delta(t_1-t_3), \\ &H^{(4)}(t_1,t_2,t_3,t_4) = H^{(3)}(t_1,t_2,t_3)\cdot H^{(1)}(t_4) \\ &\qquad - \mathrm{H}^{(2)}(t_1,t_2)\cdot\delta(t_3-t_4) \\ &\qquad - \mathrm{H}^{(2)}(t_1,t_3)\cdot\delta(t_2-t_4) \\ &\qquad - \mathrm{H}^{(2)}(t_2,t_3)\cdot\delta(t_1-t_4), \end{aligned} \tag{39}$$

In which n(t) is the white noise as noted with the following statistical properties

$$\begin{aligned} &En(t) = 0, \\ &En(t_1)\cdot n(t_2) = \delta(t_1-t_2), \end{aligned} \tag{40}$$

Where δ (-) is the Dirac delta function and. The Wiener-Hermite set is a statistically orthogonal set, i.e.

$$EH^{(i)} \cdot H^{(j)} = 0 \quad \forall\, i \neq j. \tag{41}$$

The average of almost all H functions vanishes, particularly,

$$EH^{(i)} = 0 \;\text{ for }\; i \geq 1. \tag{42}$$

Due to the completeness of the Wiener-Hermite set, any random function $G(t;\omega)$ can be expanded as

$$\begin{aligned} G(t;\omega) = {} & G^{(0)}(t) + \int_{-\infty}^{\infty} G^{(1)}(t;t_1) H^{(1)}(t_1)\,\mathrm{d}t_1 \\ & + \int_{-\infty}^{\infty}\int_{-\infty}^{\infty} G^{(2)}(t;t_1,t_2) H^{(2)}(t_1,t_2)\,\mathrm{d}t_1\mathrm{d}t_2 + \cdots \end{aligned} \tag{43}$$

Where the first two terms are the Gaussian part of $G(t;\omega)$.The rest of the terms in the expansion represent the non-Gaussian part of $G(t;\omega)$. The average of $G(t;\omega)$ is:

$$\mu_G = EG(t;\omega) = G^{(0)}(t) \tag{44}$$

The covariance of $G(t;\omega)$ is

$$
\begin{aligned}
&Cov\big(G(t;\omega),G(\tau;\omega)\big)\\
&=E\big(G(t;\omega)-\mu_G(t)\big)\big(G(\tau;\omega)-\mu_G(\tau)\big)\\
&=\int_{-\infty}^{\infty} G^{(1)}(t;t_1)G^{(1)}(\tau,t_1)\,\mathrm{d}t_1\\
&\quad+2\int_{-\infty}^{\infty}\int_{-\infty}^{\infty} G^{(2)}(t;t_1,t_2)G^{(2)}(\tau;t_1,t_2)\,\mathrm{d}t_1\mathrm{d}t_2
\end{aligned}
\tag{45}
$$

The variance of $G(t;\omega)$ is

$$
\begin{aligned}
Var\,G(t;\omega)&=E\big(G(t;\omega)-\mu_G(t)\big)^2\\
&=\int_{-\infty}^{\infty}\left[G^{(1)}(t;t_1)\right]^2\mathrm{d}t_1\\
&\quad+2\int_{-\infty}^{\infty}\int_{-\infty}^{\infty}\left[G^{(2)}(t;t_1,t_2)\right]^2\mathrm{d}t_1\mathrm{d}t_2
\end{aligned}
\tag{46}
$$

The WHEP technique can be applied on linear or nonlinear perturbed systems described by ordinary or partial differential equations. The solution can be modified in the sense that additional parts of the WienerHermite expansion can always be taken into considerations and the required order of approximations can always be made depending on the computing tool. It can be even run through a package if it is coded in some sort of symbolic languages. The technique was successfully applied to several nonlinear stochastic equations, see [20-25].

The first order solution can be obtained when considering only the Gaussian part of the solution process, i.e $u(t,x)$ can be expanded as:

$$u(t,x)=u^{(0)}(t,x)+\int_0^t u^{(1)}(t,x;x_1)H^{(1)}(x_1)\,\mathrm{d}x_1. \tag{47}$$

Substituting in the original equation (1) and taking the necessary averages, we get the following two sets of deterministic equations:

$$\begin{aligned} i)\quad \frac{\partial u^{(0)}(t,x)}{\partial t} &= \frac{\partial^2 u^{(0)}}{\partial x^2}-\varepsilon\left[u^{(0)}(t,x)\right]^2 \\ &\quad -3\varepsilon u^{(0)}(t,x)\int_0^x\left[u^{(1)}(t,x;x_1)\right]^2\,\mathrm{d}x_1, \\ u^{(0)}(t,0)=0,\ u^{(0)}(t,L)&=0 \text{ and } u^{(0)}(0,x)=\varphi(x), \end{aligned} \tag{48}$$

$$\begin{aligned} ii)\quad &\frac{\partial u^{(1)}(t,x;x_1)}{\partial t} \\ &=\frac{\partial^2 u^{(1)}(t,x;x_1)}{\partial x^2}-3\varepsilon\left[u^{(0)}(t,x)\right]^2 u^{(1)}(t,x;x_1) \\ &\quad -3\varepsilon u^{(1)}(t,x;x_1)\int_0^x\left[u^{(1)}(t,x;x_1)\right]^2\,\mathrm{d}x_1+\sigma\delta(x-x_1), \\ &u^{(1)}(t,0;x_1)=0,\ u^{(1)}(t,L;x_1)=0 \text{ and } u^{(1)}(0,x;x_1)=0. \end{aligned} \tag{49}$$

Applying WHEP technique, the deterministic kernels can be represented in first order approximation as:

$$u^{(0)}=u_0^{(0)}+\varepsilon\cdot u_1^{(0)}, \tag{50}$$

$$u^{(1)}=u_0^{(1)}+\varepsilon\cdot u_1^{(1)}. \tag{51}$$

Substituting in the previous set of equations (48) and (49), we get the following four sets of equations:

$$\frac{\partial u_0^{(0)}(t,x)}{\partial t}=\frac{\partial^2 u_0^{(0)}}{\partial x^2}$$
$$u_0^{(0)}(t,0)=0, u_0^{(0)}(t,L)=0 \text{ and } u_0^{(0)}(0,x)=\varphi(x), \tag{52}$$

$$\frac{\partial u_1^{(0)}(t,x)}{\partial t}=\frac{\partial^2 u_1^{(0)}(t,x)}{\partial x^2}-\left[u_0^{(0)}(t,x)\right]^3$$
$$-3u_0^{(0)}(t,x)\int_0^x\left[u_0^{(1)}(t,x)\right]^2 dx_1,$$
$$u_1^{(0)}(t,0)=0, u_1^{(0)}(t,L)=0 \text{ and } u_1^{(0)}(0,x)=0, \tag{53}$$

$$\frac{\partial u_0^{(1)}(t,x;x_1)}{\partial t}=\frac{\partial^2 u_0^{(1)}(t,x;x_1)}{\partial x^2}+\sigma\cdot\delta(x-x_1)$$
$$u_0^{(1)}(t,0;x_1)=0, u_0^{(1)}(t,L;x_1)=0 \text{ and } u_0^{(1)}(0,x;x_1)=0, \tag{54}$$

$$\frac{\partial u_1^{(1)}(t,x;x_1)}{\partial t}=\frac{\partial^2 u_1^{(1)}(t,x;x_1)}{\partial x^2}$$
$$-3\left[u_0^{(0)}(t,x)\right]^2\cdot u_0^{(1)}(t,x;x_1)$$
$$-3u_0^{(1)}(t,x;x_1)\int_0^x\left[u_0^{(1)}(t,x;x_1)\right]^2 dx_1$$
$$u_1^{(1)}(t,0;x_1)=0, u_1^{(1)}(t,L;x_1)=0 \text{ and } u_1^{(1)}(0,x;x_1)=0. \tag{55}$$

The algorithm of solution is evaluating $u_0^{(0)}$ and $u_0^{(1)}$ first using the separation of variables and the Eigen function expansion respectively and then computing the other two kernels independently using the Eigenfunction expansion. The final results are:

$$Eu(x,t) = u_0^{(0)}(x,t) + \varepsilon \cdot u_1^{(0)}(x,t), \tag{56}$$

$$\begin{aligned} Var\ u(t,x) &= \int_0^x \left[u_0^{(1)}(t,x;x_1)\right]^2 dx_1 \\ &+ 2\varepsilon \int_0^x u_0^{(1)}(t,x;x_1) u_1^{(0)}(t,x;x_1) dx_1 \\ &+ \varepsilon^2 \int_0^x \left[u_1^{(1)}(t,x;x_1)\right]^2 dx_1 \end{aligned} \tag{57}$$

The result can be made better using the second correction as the following formula:

$$u^{(0)} = u_0^{(0)} + \varepsilon \cdot u_1^{(0)}, \tag{58}$$

$$u^{(1)} = u_0^{(1)} + \varepsilon_1^{(1)} u_1^{(1)}. \tag{59}$$

The following are some sample results.

2.2.1. First Order Approximation and Different Corrections Results

Figures 7 and 8 illustrate the change of the first order mean and variance under different correction levels.

We can notice the stable results for the different corrections.

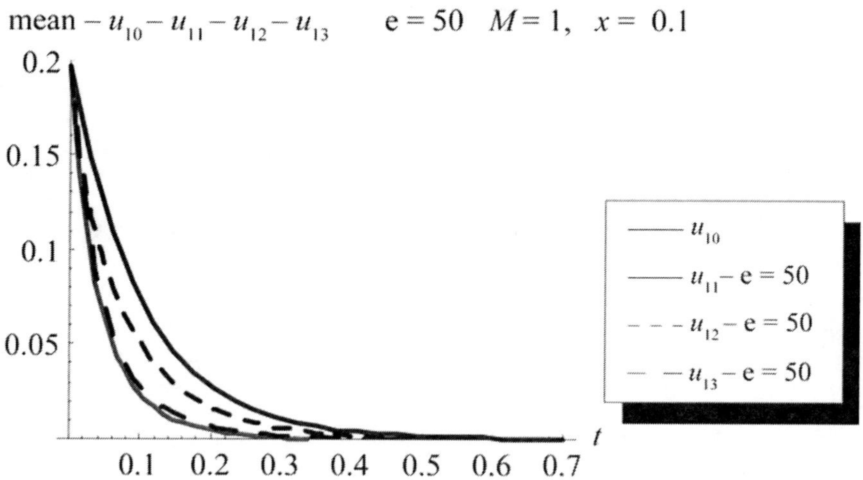

Figure 7. Mean comparison between first order with zero, first, second and third corrections at ε = 50 at x = 0.1, (L = 1, M = 1, σ = 1, Φ(x) = x).

2.2.2. Second Order Approximation and Different Corrections Results

Figures 9 and 10 illustrate the change of the second order mean and variance under different correction levels.

2.2.3. Mean and Variance Comparisons between First and Second Order Approximations with Different Corrections Results

Figures 11 and 12 illustrate in a comparative way, the mean and variance under first and second orders with some correction levels.

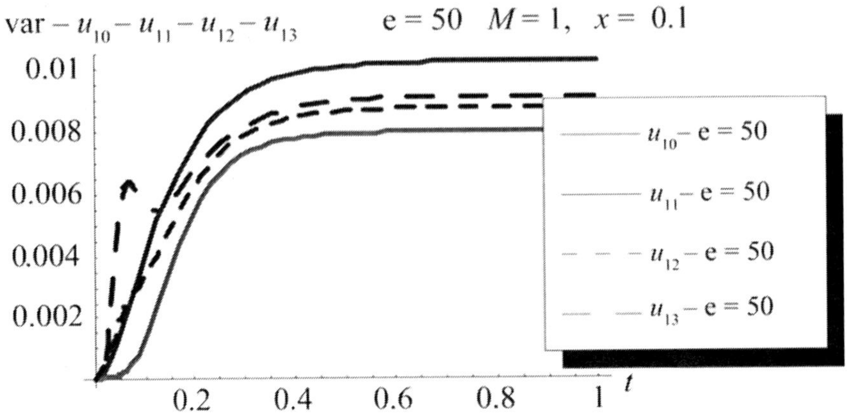

Figure 8. Variance comparison between first order with zero, first, second and third corrections at x = 0.1 at ε = 50, (L = 1, M = 1, σ = 1, Φ(x) = x).

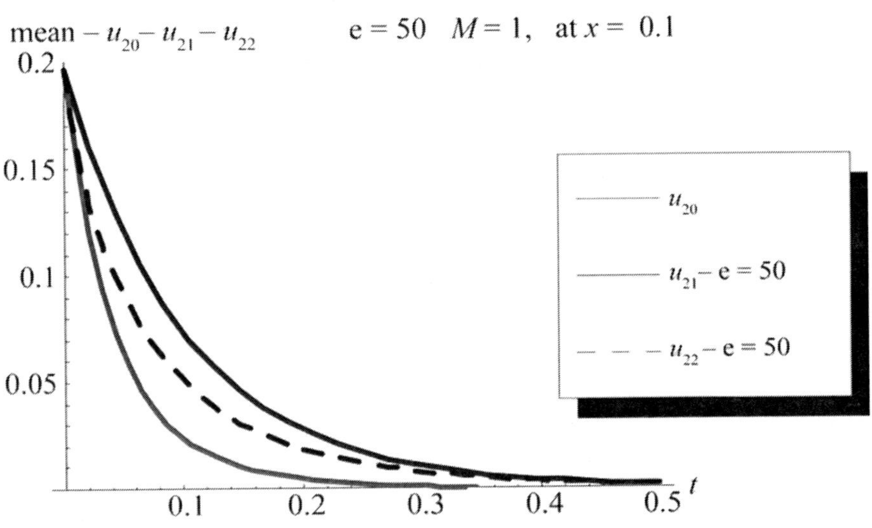

Figure 9. Mean comparison between second order with zero, first and second corrections at x = 0.1, ε = 50, (L = 1, M = 1, σ = 1, Φ(x) = x).

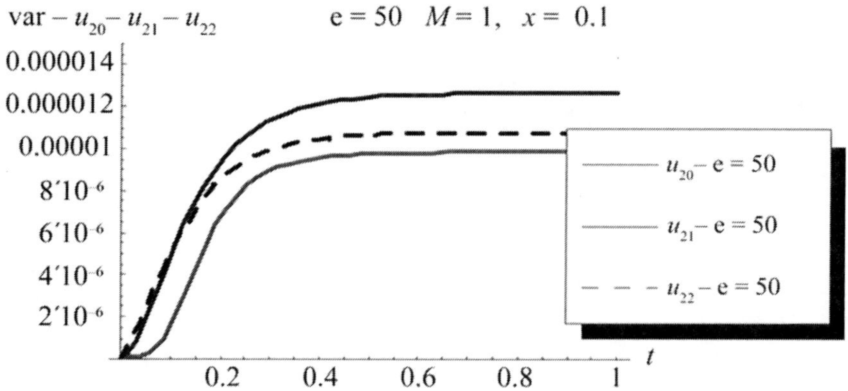

Figure 10. Variance comparison between second order with zero, first and second corrections at x = 0.1, ε = 50, (L = 1, M = 1, σ = 1, Φ(x) = x).

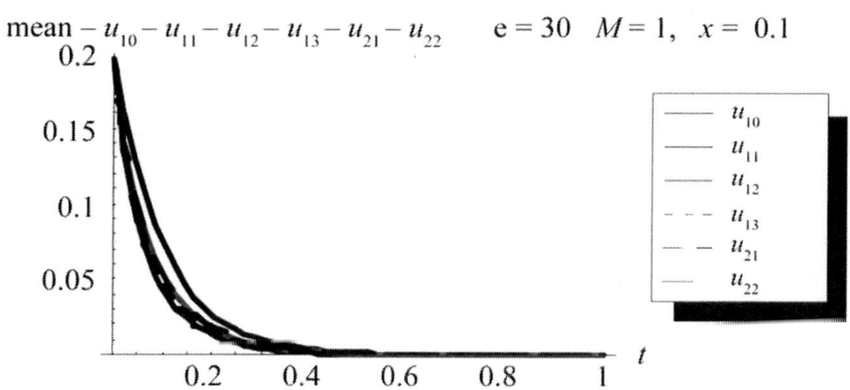

Figure 11. Mean comparison between first order with zero, first, second and third corrections, second order with first and second corrections at x = 0.1 at ε = 30, (L = 1, M = 1, σ = 1, Φ(x) = x).

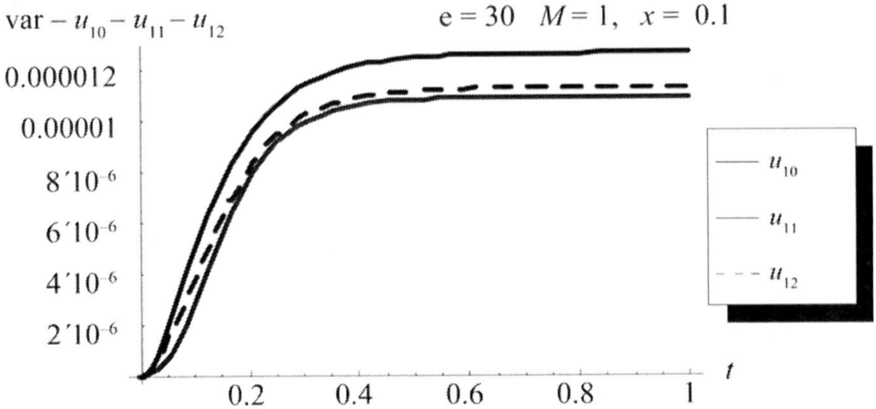

Figure 12. Variance comparison between first order with zero, first and second corrections at x = 0.1 at ε = 30 (L = 1, M = 1, σ = 1, Ψ(x) = x).

2.3. Using Homotopy Perturbation Technique (HPM)

In homotopy Perturbation method (HPM) [18-21], a parameter $p \in [0,1]$ is embedded in a homotopy function $v(r,p):\phi\times[0,1]\to\Re$ which satisfies

$$H(v,p)=\left[L(v)-L(u_0)\right]+p\left[A(v)-f(r)\right]=0 \tag{60}$$

Where u_0 is an initial approximation to the solution of the equation:

$$A(u)-f(r)=0, r\in\phi \tag{61}$$

With boundary condition

$$B\left(u,\frac{\partial u}{\partial n}\right)=0, r\in\phi \tag{62}$$

In which A is a nonlinear differential operator which can be decomposed into a linear operator R and anon linear operator N,B is a boundary operator, f(r) is a known analytic function and Γ is the boundary of Φ .the homotopy introduces a continuously deformed solution for the case $p=0$, $R(v)-L(u_0)=0$, which is the original equation .this is the basic idea of the homotopy method which is to deform continuously a simple problem (and easy to solve)into the difficult problem under study.

The basic assumption of the (HPM) method is that the solution of the original equation can be expanded as a power series in p as:

$$v = v_0 + pv_1 + p^2 v_2 + p^3 v_3 + \cdots \tag{63}$$

Now, setting p = 1, the approximation solution is obtained as:

$$u = \lim_{p \to 1} v = v_0 + v_1 + v_2 + v_3 + \cdots \tag{64}$$

The rate of convergence of the method depends greatly on the initial approximation u_0 which is considered as the main disadvantage of HPM. The idea of imbedded parameter can be utilized to solve nonlinear problems by imbedding this parameter to the problem and then forcing it to be unity in the obtained approximate solution if convergence can be assured .it is a simple technique which enables the extension of the applicability of the perturbation method from small value applications to general ones.

Applying HPM on equation (1), we can get the following results w. r. t homotopy perturbation:

$$A(u)=R(u)+\varepsilon u^3 \tag{65}$$

$$R(u)=\frac{\partial u(t,x;\omega)}{\partial t}-\frac{\partial^2 u}{\partial x^2} \tag{66}$$

$$N(u)=\varepsilon u^3 \tag{67}$$

$$f(r)=\sigma\cdot n(x;\omega) \tag{68}$$

The homotopy function takes the following form:

$$H(\omega,p)=\left[R(\omega)-R(z_0)\right]+p\left[A(\omega)-f(r)\right]=0 \tag{69}$$

Or equivalently,

$$R(\omega)-R(z_0)+p\left[R(z_0)+\varepsilon\omega^3-\sigma\cdot n(x;\omega)\right]=0 \tag{70}$$

Where z_0 is an initial solution .The approximate solution can then be obtained using

$$\omega=\omega_0+p\omega_1+p^2\omega_2+p^3\omega_3+\cdots \tag{71}$$

Now, setting p = 1, the approximation solution is obtained as:

$$u = \lim_{p \to 1} \omega = \omega_0 + \omega_1 + \omega_2 + \omega_3 + \cdots \tag{72}$$

Using equation (86) in equation (85) and equating the equal powers of p in both sides of the equation, one can get the following results:

i) $R(\omega_0) = R(z_0)$, in which one may consider the following simple solution:

$$z_0(t,0) = z_0(t,L) = 0,\ \omega_0 = z_0,$$

$$\text{and } z_0(0,x) = \phi(x) \tag{73}$$

$$ii)\ R(\omega_1) = \sigma \cdot n(x;\omega) - R(\omega_0) - \varepsilon\omega_0^3$$
$$\omega_1(t,0) = \omega_1(t,L) = 0, \text{ and } \omega_1(0,x) = 0 \tag{74}$$

$$iii)\ R(\omega_2) = -3\varepsilon\omega_0^2\omega_1$$
$$\omega_2(t,0) = \omega_2(t,L) = 0, \text{ and } \omega_2(0,x) = 0 \tag{75}$$

$$iv)\ R(\omega_3) = -3\varepsilon\left(\omega_0\omega_1^2 + \omega_0^2\omega_2\right)$$
$$\omega_3(t,0) = \omega_3(t,L) = 0, \text{ and } \omega_3(0,x) = 0 \tag{76}$$

$$vi)\ R(\omega_4) = -\varepsilon\left(\omega_1^3 + 6\omega_0\omega_1\omega_2 + 3\omega_0^2\omega_2\right)$$
$$\omega_4(t,0) = \omega_4(t,L) = 0, \text{ and } \omega_4(0,x) = 0 \tag{77}$$

As before we have many choices in guessing the initial approximation together with its initial conditions which greatly affects the consequent approximation .the choice ω_0 is a design problem which can be taken as follows:

$$\omega_0(t,x)=\sum_{n=0}^{\infty}A_n e^{\beta_n t}\sin\frac{n\pi}{L}x$$
$$A_n=\int_0^L\phi(x)\sin\frac{n\pi}{L}x\,dx \tag{78}$$

One can notice that the selected value function satisfies the initial and boundary conditions and it depends on the parameter β_n which is totally free .One can also notice that β_n selection could control the solution convergence.

The first order approximation can be obtained using Eigen function expansion as follows:

$$u^{(1)}(t,x;\omega)=\omega_0+\omega_1=\omega_0+\sum_{n=0}^{\infty}j_{n_1}(t)\sin\frac{n\pi}{L}x$$

where

$$j_{n_1}(t)=\int_0^t e^{\left(\frac{-n\pi}{L}\right)^2(t-\tau)}G_{n_1}(\tau)\,d\tau$$
$$G_{n_1}(t)=\frac{2}{L}\int_0^L\left[\sigma\cdot n(x;\omega)-R(\omega_0)-\varepsilon\omega_0^3\right]dx \tag{79}$$

The ensemble average is

$$\mu_{u^{(1)}}(t,x;\omega)=\omega_0+\sum_{n=0}^{\infty}Ej_{n_1}(t)\sin\frac{n\pi}{L}x$$

where

$$Ej_{n_1}(t)=\int_0^t e^{\left(\frac{-n\pi}{L}\right)^2(t-\tau)}EG_{n_1}(\tau)\,d\tau$$
$$EG_{n_1}(t)=\frac{-2}{L}\int_0^L\left[R(\omega_0)+\varepsilon\omega_0^3\right]\sin\frac{n\pi}{L}x\,dx \tag{80}$$

The covariance is obtained from the following final expression

$$Cov\left(u^{(1)}(t,x_1),u^{(1)}(t,x_2)\right)$$
$$=\frac{4\sigma^2}{L^2}\sum_{n=1}^{\infty}\sum_{m=1}^{\infty}\sin\frac{n\pi}{L}x_1\sin\frac{m\pi}{L}x_2\left(\int_0^L\sin\frac{n\pi}{L}x\sin\frac{m\pi}{L}x\,dx\right)$$
$$\left(\int_0^t\int_0^t e^{\left(\frac{-n\pi}{L}\right)^2(t-\tau_1)}e^{\left(\frac{-m\pi}{L}\right)^2(t-\tau_2)}d\tau_1 d\tau_2\right) \quad (81)$$

The variance can then be obtained from equation (81) by setting $x_1 = x_2 = x$.

Any higher order approximations can be obtained in a similar way. The following sample results using the same data in the cubic case.

Computing the consequent errors Er_i by using the following expression

$$Er_{i-1,i}=\left|mean_i-mean_{i-1}\right|,i=2,3,4 \quad (82)$$

we obtain the following results. Figures 13 illustrates different mean approximations using homotopy method (HPM) with computing their corresponding decreasing errors in figure 14.

Figures 15 illustrates different variance approximations using homotopy method (HPM) with computing their corresponding decreasing errors in figure 16.

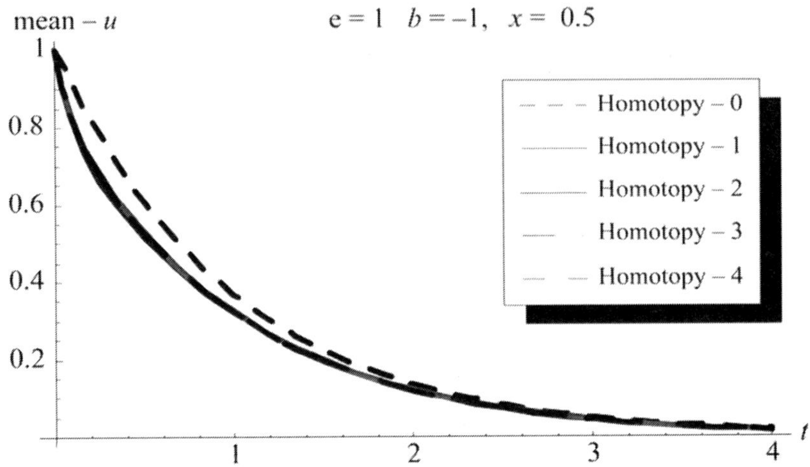

Figure 13. The mean at different homotopy orders, cubic case.

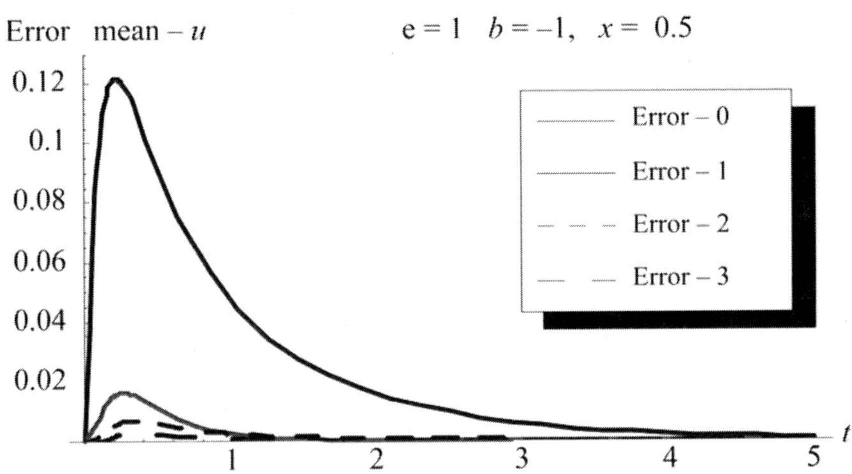

Figure 14. The error differences, cubic case.

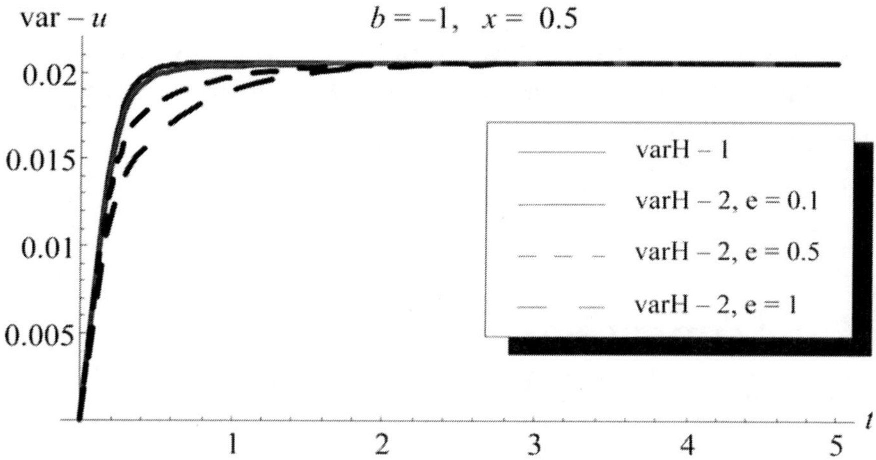

Figure 15. The variance (-1: first and-2: second) at different ε, HPM, cubic nonlinear case.

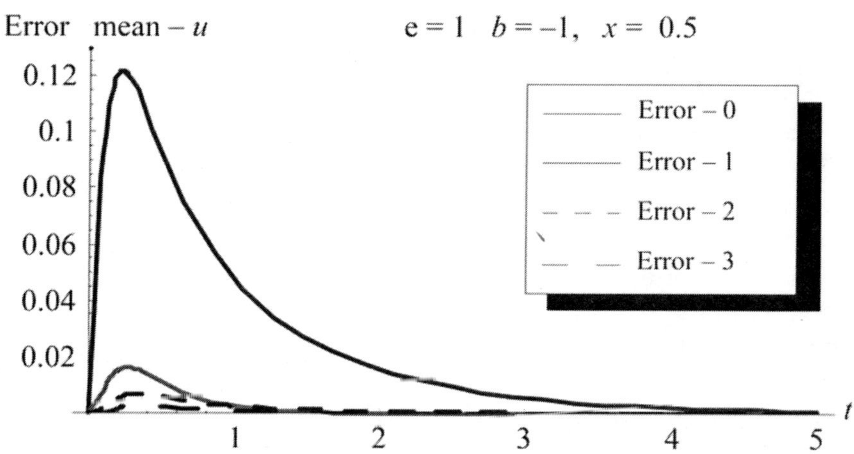

Figure 16. The error difference of first and second error at different ε, cubic case.

2.4. Mean and Variance Comparisons of the Solution of Stochastic Cubic Nonlinear Diffusion Problem Using WHEP, Pickard and HPM Methods with Different Orders and Corrections

Figures 17, 18, 19 and 20 illustrate some useful comparisons among the three used methods in this paper.

3. THE QUINTIC NONLINEAR STOCHASTIC DIFFUSION EQUATION

Let us consider the following stochastic nonlinear-diffusion equation

$$\begin{aligned}&\frac{\partial u(t,x;\omega)}{\partial t}=\frac{\partial^2 u}{\partial x^2}-\varepsilon\cdot u^5+\sigma\cdot n(x;\omega);\\&(t,x)\in(0,\infty)\times(0,L),\ u(t,0)=0,\ u(t,L)=0\\&\text{and } u(0,x)=\varphi(x).\end{aligned} \tag{83}$$

Where ε is a deterministic scale for the nonlinear term. Two methods are used in the next subsections, mainly the Pickard approximations and the HPM technique.

3.1. Using the Pickard Approximation

The following results for the first order approximation are obtained

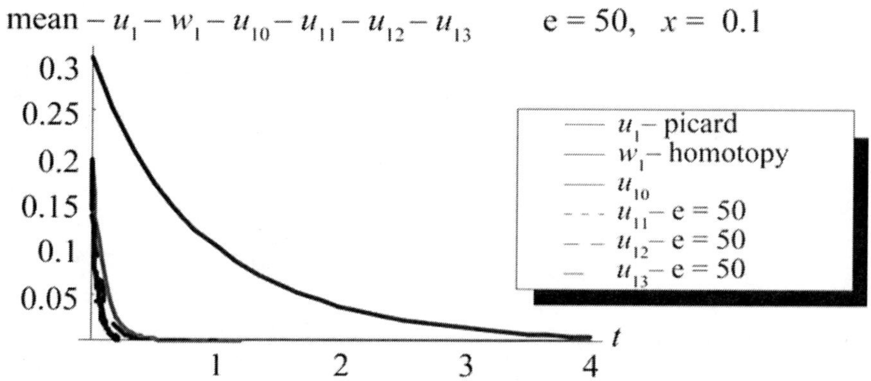

Figure 17. Mean comparison between Picard first order, Homotopy first order and WHEP first order with zero, first, second and third corrections at x = 0.1, ε = 50, (L = 1, M = 1, σ = 1, Φ(x) = x).

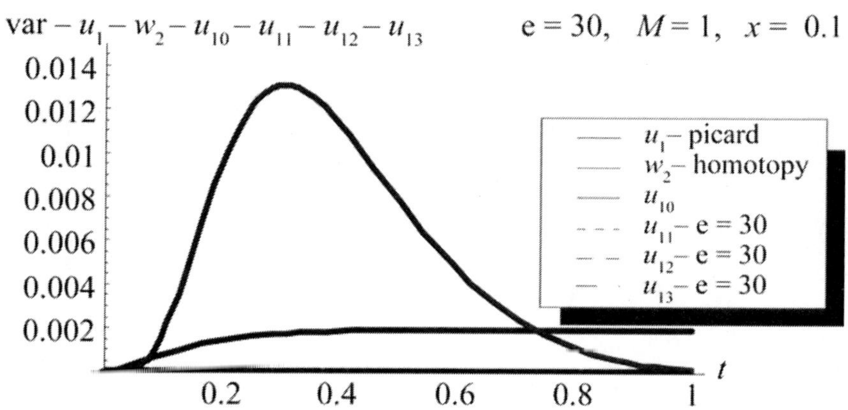

Figure 18. Variance comparison between Picard first order, Homotopy second order and WHEP first order with zero, first, second and third corrections at x = 0.1, ε = 30, (L = 1, M = 1, σ = 1, Φ(x) = x).

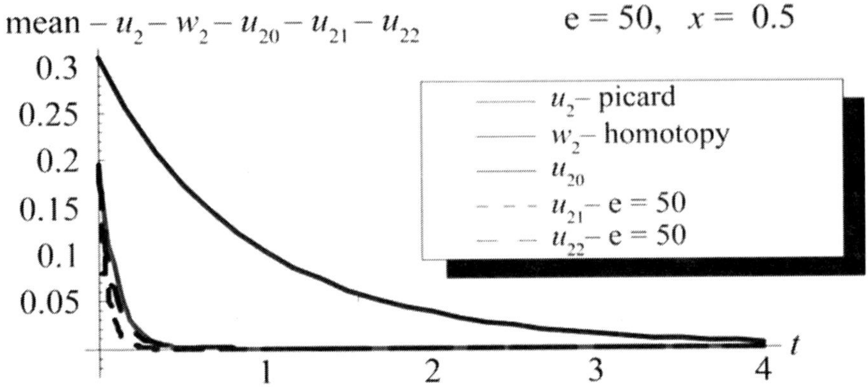

Figure 19. Mean comparison between Picard second order , Homotopy fourth order and WHEP second order with zero, first, second and third corrections at x = 0.5 , ε = 50, (L = 1, M = 1, σ = 1, Φ(x) = x).

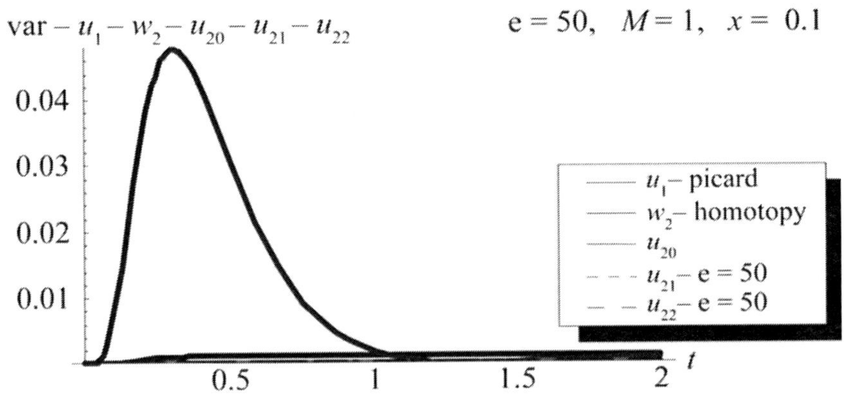

Figure 20. Variance comparison between Picard first order, Homotopy second order and WHEP second order with zero, first and second corrections at x = 0.1, ε = 50, (L = 1, M = 1, σ = 1, Φ(x) = x).

$$u_1(t,x)=\sum_{n=0}^{\infty}T_{n_0}e^{-\left(\frac{n\pi}{L}\right)^2 t}\sin\frac{n\pi}{L}x+\sum_{n=0}^{\infty}\left(\int_0^t e^{-\left(\frac{n\pi}{L}\right)^2(t-\tau)}F_{n_1}(\tau)d\tau\right)\sin\frac{n\pi}{L}x \tag{84}$$

$$F_{n_1}(t)=\frac{2\sigma}{L}\int_0^L n(x)\sin\frac{n\pi}{L}x-\frac{2\varepsilon}{L}\int_0^L\left[u_0(t,x)\right]^3\sin\frac{n\pi}{L}x\,dx \tag{85}$$

where

$$\begin{aligned}
&\left[u_0(t,x)\right]^5\\
&=\mu_{u_0}^5(t,x)+5\mu_{u_0}^4(t,x)\sum_{n=0}^{\infty}I_{n_0}(t)\sin\frac{n\pi}{L}x\\
&\quad+10\mu_{u_0}^3(t,x)\sum_{n=0}^{\infty}\sum_{m=0}^{\infty}I_{n_0}(t)I_{m_0}(t)\sin\frac{n\pi}{L}x\sin\frac{m\pi}{L}x\\
&\quad+10\mu_{u_0}^2(t,x)\sum_{n=0}^{\infty}\sum_{m=0}^{\infty}\sum_{l=0}^{\infty}I_{n_0}(t)I_{m_0}(t)I_{l_0}(t)\sin\frac{n\pi}{L}x\\
&\quad\sin\frac{m\pi}{L}x\sin\frac{l\pi}{L}x\\
&\quad+5\mu_{u_0}(t,x)\sum_{n=0}^{\infty}\sum_{m=0}^{\infty}\sum_{l=0}^{\infty}\sum_{k=0}^{\infty}I_{n_0}(t)I_{m_0}(t)I_{l_0}(t)I_{k_0}(t)\\
&\quad\sin\frac{n\pi}{L}x\sin\frac{m\pi}{L}x\sin\frac{l\pi}{L}x\sin\frac{k\pi}{L}x\\
&\quad+\sum_{n=0}^{\infty}\sum_{m=0}^{\infty}\sum_{l=0}^{\infty}\sum_{k=0}^{\infty}\sum_{v=0}^{\infty}I_{n_0}(t)I_{m_0}(t)I_{l_0}(t)I_{k_0}(t)I_{v_0}(t)\\
&\quad\sin\frac{n\pi}{L}x\sin\frac{m\pi}{L}x\sin\frac{l\pi}{L}x\sin\frac{k\pi}{L}x\sin\frac{v\pi}{L}x
\end{aligned} \tag{86}$$

First order mean

$$\mu_{u_1}(t,x)=\sum_{n=0}^{\infty}T_{n_0}e^{-\left(\frac{n\pi}{L}\right)^2 t}\sin\frac{n\pi}{L}x$$
$$+\sum_{n=0}^{\infty}\left(\int_0^t e^{-\left(\frac{n\pi}{L}\right)^2(t-\tau)}\mu_{F_{n_1}}(\tau)\mathrm{d}\tau\right)\sin\frac{n\pi}{L}x \tag{87}$$

where

$$\mu_{F_{n_1}}(t)=\frac{-2\varepsilon}{L}\int_0^L E\left[u_0(t,x)\right]^5\sin\frac{n\pi}{L}x\,\mathrm{d}x \tag{88}$$

$$E\left[u_0(t,x)\right]^5$$
$$=\mu_{u_0}^5(t,x)+10\mu_{u_0}^3(t,x)\sigma_{u_0}^2(t,x)+5\mu_{u_0}(t,x)$$
$$\sum_{n=0}^{\infty}\sum_{m=0}^{\infty}\sum_{l=0}^{\infty}\sum_{k=0}^{\infty}EI_{n_0}(t)I_{m_0}(t)I_{l_0}(t)I_{k_0}(t)$$
$$\sin\frac{n\pi}{L}x\sin\frac{m\pi}{L}x\sin\frac{l\pi}{L}x\sin\frac{k\pi}{L}x \tag{89}$$

$$Cov(u_1(t,x_1),u_1(t,x_2))$$
$$=\sum_{n=0}^{\infty}\sum_{m=0}^{\infty}EI_{n_1}(t)I_{m_1}(t)\sin\frac{n\pi}{L}x_1\sin\frac{m\pi}{L}x_2$$
$$-\sum_{n=0}^{\infty}\left(\int_0^t e^{-\left(\frac{n\pi}{L}\right)^2(t-\tau)}EF_{n_1}(\tau)\mathrm{d}\tau\right)\sin\frac{n\pi}{L}x_2$$
$$\cdot\sum_{n=0}^{\infty}EI_{n_1}(t)\sin\frac{n\pi}{L}x_1-\sum_{n=0}^{\infty}\left(\int_0^t e^{-\left(\frac{n\pi}{L}\right)^2(t-\tau)}EF_{n_1}(\tau)\mathrm{d}\tau\right)$$
$$\sin\frac{n\pi}{L}x_1\cdot\sum_{n=0}^{\infty}EI_{n_1}(t)\sin\frac{n\pi}{L}x_2$$
$$+\sum_{n=0}^{\infty}EI_{n_1}(t)\sin\frac{n\pi}{L}x_1\cdot\sum_{n=0}^{\infty}EI_{n_1}(t)\sin\frac{n\pi}{L}x_2 \tag{90}$$

where

$$
\begin{aligned}
&En(x)u_0^5(t,y)\\
&=5\left(\sum_{n=0}^{\infty}T_{n_0}\,e^{-\left(\frac{n\pi}{L}\right)^2 t}\sin\frac{n\pi}{L}y\right)^5\sum_{n=0}^{\infty}\sin\frac{n\pi}{L}y\,En(x)I_{n_0}(t)\\
&\quad+10\left(\sum_{n=0}^{\infty}T_{n_0}\,e^{-\left(\frac{n\pi}{L}\right)^2 t}\sin\frac{n\pi}{L}y\right)^2\sum_{n=0}^{\infty}\sum_{m=0}^{\infty}\sum_{l=0}^{\infty}En(x)I_{n_0}(t)\\
&\qquad I_{m_0}(t)I_{l_0}(t)\sin\frac{n\pi}{L}y\sin\frac{m\pi}{L}y\sin\frac{l\pi}{L}y\\
&\quad+\sum_{n=0}^{\infty}\sum_{m=0}^{\infty}\sum_{l=0}^{\infty}\sum_{k=0}^{\infty}\sum_{v=0}^{\infty}En(x)I_{n_0}(t)I_{m_0}(t)I_{l_0}(t)\\
&\qquad I_{k_0}(t)I_{v_0}(t)\sin\frac{n\pi}{L}y\sin\frac{m\pi}{L}y\sin\frac{l\pi}{L}y\\
&\qquad\sin\frac{k\pi}{L}y\sin\frac{v\pi}{L}y
\end{aligned}
\tag{91}
$$

$$
\begin{aligned}
&Eu_0^5(\tau_1,x_3)u_0^5(\tau_2,x_4)\\
&=\phi^5[\tau_1,x_3]\phi^5[\tau_2,x_4]+10\phi^5[\tau_1,x_3]\phi^3[\tau_2,x_4]\sum_{n=0}^{\infty}\sum_{m=0}^{\infty}EI_{n_0}(\tau_2)I_{m_0}(\tau_2)\sin\frac{n\pi}{l}x_4\sin\frac{m\pi}{l}x_4\\
&\quad+10\phi^5[\tau_2,x_4]\phi^3[\tau_1,x_3]\sum_{n=0}^{\infty}\sum_{m=0}^{\infty}EI_{n_0}(\tau_1)I_{m_0}(\tau_1)\sin\frac{n\pi}{l}x_3\sin\frac{m\pi}{l}x_3\\
&\quad+25\phi^4[\tau_1,x_3]\phi^3[\tau_2,x_4]\sum_{n=0}^{\infty}\sum_{m=0}^{\infty}EI_{n_0}(\tau_1)I_{m_0}(\tau_2)\sin\frac{n\pi}{l}x_3\sin\frac{m\pi}{l}x_4
\end{aligned}
$$

$$
\begin{aligned}
&+5\phi[\tau_1,x_3]\phi[\tau_2,x_4]\sum_{n=0}^{\infty}\sum_{m=0}^{\infty}\sum_{l=0}^{\infty}\sum_{k=0}^{\infty}EI_{n_0}(\tau_2)I_{m_0}(\tau_2)I_{l_0}(\tau_2)I_{k_0}(\tau_2)\sin\frac{n\pi}{l}x_4\sin\frac{m\pi}{l}x_4\sin\frac{l\pi}{l}x_4\sin\frac{k\pi}{l}x_4\\
&+50\phi^4[\tau_1,x_3]\phi^2[\tau_2,x_4]\sum_{n=0}^{\infty}\sum_{m=0}^{\infty}\sum_{l=0}^{\infty}\sum_{k=0}^{\infty}EI_{n_0}(\tau_1)I_{m_0}(\tau_2)I_{l_0}(\tau_2)I_{k_0}(\tau_2)\sin\frac{n\pi}{l}x_3\sin\frac{m\pi}{l}x_4\sin\frac{l\pi}{l}x_4\sin\frac{k\pi}{l}x_4\\
&+50\phi^2[\tau_1,x_3]\phi^4[\tau_2,x_4]\sum_{n=0}^{\infty}\sum_{m=0}^{\infty}\sum_{l=0}^{\infty}\sum_{k=0}^{\infty}EI_{n_0}(\tau_1)I_{m_0}(\tau_1)I_{l_0}(\tau_1)I_{k_0}(\tau_2)\sin\frac{n\pi}{l}x_3\sin\frac{m\pi}{l}x_3\sin\frac{l\pi}{l}x_3\sin\frac{k\pi}{l}x_4\\
&+50\phi^4[\tau_1,x_3]\phi^2[\tau_2,x_4]\sum_{n=0}^{\infty}\sum_{m=0}^{\infty}\sum_{l=0}^{\infty}\sum_{k=0}^{\infty}EI_{n_0}(\tau_1)I_{m_0}(\tau_2)I_{l_0}(\tau_2)I_{k_0}(\tau_2)\sin\frac{n\pi}{l}x_3\sin\frac{m\pi}{l}x_4\sin\frac{l\pi}{l}x_4\sin\frac{k\pi}{l}x_4\\
&+100\phi^3[\tau_1,x_3]\phi^3[\tau_2,x_4]\sum_{n=0}^{\infty}\sum_{m=0}^{\infty}\sum_{l=0}^{\infty}\sum_{k=0}^{\infty}EI_{n_0}(\tau_1)I_{m_0}(\tau_1)I_{l_0}(\tau_2)I_{k_0}(\tau_2)\sin\frac{n\pi}{l}x_3\sin\frac{m\pi}{l}x_3\sin\frac{l\pi}{l}x_4\sin\frac{k\pi}{l}x_4\\
&+5\phi^4[\tau_1,x_3]\sum_{n=0}^{\infty}\sum_{m=0}^{\infty}\sum_{l=0}^{\infty}\sum_{k=0}^{\infty}\sum_{v=0}^{\infty}\sum_{w=0}^{\infty}EI_{n_0}(\tau_1)I_{m_0}(\tau_2)I_{l_0}(\tau_2)I_{k_0}(\tau_2)I_{v_0}(\tau_2)I_{w_0}(\tau_2)\\
&\quad\sin\frac{n\pi}{l}x_3\sin\frac{m\pi}{l}x_4\sin\frac{l\pi}{l}x_4\sin\frac{k\pi}{l}x_4\sin\frac{v\pi}{l}x_4\sin\frac{w\pi}{l}x_4\\
&+50\phi^3[\tau_1,x_3]\phi[\tau_2,x_4]\sum_{n=0}^{\infty}\sum_{m=0}^{\infty}\sum_{l=0}^{\infty}\sum_{k=0}^{\infty}\sum_{v=0}^{\infty}\sum_{w=0}^{\infty}EI_{n_0}(\tau_1)I_{m_0}(\tau_1)I_{l_0}(\tau_2)I_{k_0}(\tau_2)I_{v_0}(\tau_2)I_{w_0}(\tau_2)\\
&\quad\sin\frac{n\pi}{l}x_3\sin\frac{m\pi}{l}x_3\sin\frac{l\pi}{l}x_4\sin\frac{k\pi}{l}x_4\sin\frac{v\pi}{l}x_4\sin\frac{w\pi}{l}x_4\\
&+100\phi^2[\tau_1,x_3]\phi^2[\tau_2,x_4]\sum_{n=0}^{\infty}\sum_{m=0}^{\infty}\sum_{l=0}^{\infty}\sum_{k=0}^{\infty}\sum_{v=0}^{\infty}\sum_{w=0}^{\infty}EI_{n_0}(\tau_1)I_{m_0}(\tau_1)I_{l_0}(\tau_1)I_{k_0}(\tau_2)I_{v_0}(\tau_2)I_{w_0}(\tau_2)\\
&\quad\sin\frac{n\pi}{l}x_3\sin\frac{m\pi}{l}x_3\sin\frac{l\pi}{l}x_3\sin\frac{k\pi}{l}x_4\sin\frac{v\pi}{l}x_4\sin\frac{w\pi}{l}x_4\\
&+25\phi[\tau_1,x_3]\phi[\tau_2,x_4]\sum_{n=0}^{\infty}\sum_{m=0}^{\infty}\sum_{l=0}^{\infty}\sum_{k=0}^{\infty}\sum_{v=0}^{\infty}\sum_{w=0}^{\infty}\sum_{z=0}^{\infty}\sum_{q=0}^{\infty}EI_{n_0}(\tau_1)I_{m_0}(\tau_1)I_{l_0}(\tau_1)I_{k_0}(\tau_1)I_{v_0}(\tau_2)I_{w_0}(\tau_2)I_{z_0}(\tau_2)I_{q_0}(\tau_2)\\
&\quad\sin\frac{n\pi}{l}x_3\sin\frac{m\pi}{l}x_3\sin\frac{l\pi}{l}x_3\sin\frac{k\pi}{l}x_3\sin\frac{v\pi}{l}x_4\sin\frac{w\pi}{l}x_4\sin\frac{z\pi}{l}x_4\sin\frac{q\pi}{l}x_4\\
&+10\phi^2[\tau_2,x_4]\sum_{n=0}^{\infty}\sum_{m=0}^{\infty}\sum_{l=0}^{\infty}\sum_{k=0}^{\infty}\sum_{v=0}^{\infty}\sum_{w=0}^{\infty}\sum_{z=0}^{\infty}\sum_{q=0}^{\infty}EI_{n_0}(\tau_1)I_{m_0}(\tau_1)I_{l_0}(\tau_1)I_{k_0}(\tau_1)I_{v_0}(\tau_1)I_{w_0}(\tau_2)I_{z_0}(\tau_2)I_{q_0}(\tau_2)\\
&\quad\sin\frac{n\pi}{l}x_3\sin\frac{m\pi}{l}x_3\sin\frac{l\pi}{l}x_3\sin\frac{k\pi}{l}x_3\sin\frac{v\pi}{l}x_3\sin\frac{w\pi}{l}x_4\sin\frac{z\pi}{l}x_4\sin\frac{q\pi}{l}x_4\\
&+\sum_{n=0}^{\infty}\sum_{m=0}^{\infty}\sum_{l=0}^{\infty}\sum_{k=0}^{\infty}\sum_{v=0}^{\infty}\sum_{w=0}^{\infty}\sum_{z=0}^{\infty}\sum_{q=0}^{\infty}I_{n_0}(\tau_1)I_{m_0}(\tau_1)I_{l_0}(\tau_1)I_{k_0}(\tau_1)I_{v_0}(\tau_1)I_{w_0}(\tau_2)I_{z_0}(\tau_2)I_{q_0}(\tau_2)I_{h_0}(\tau_2)I_{j_0}(\tau_2)\\
&\quad\sin\frac{n\pi}{l}x_3\sin\frac{m\pi}{l}x_3\sin\frac{l\pi}{l}x_3\sin\frac{k\pi}{l}x_3\sin\frac{v\pi}{l}x_3\sin\frac{w\pi}{l}x_4\sin\frac{z\pi}{l}\pi x_4\sin\frac{q\pi}{l}x_4\sin\frac{h\pi}{l}x_4\sin\frac{j\pi}{l}x_4
\end{aligned}
\tag{92}
$$

In which

$$
\begin{aligned}
&EI_{n_0}(\tau_1)I_{m_0}(\tau_1)I_{l_0}(\tau_1)I_{k_0}(\tau_1)I_{v_0}(\tau_1)I_{w_0}(\tau_1)\\
&=\int_0^{\tau_1}\int_0^{\tau_1}\int_0^{\tau_1}\int_0^{\tau_1}\int_0^{\tau_1}\int_0^{\tau_1}e^{-\left(\frac{n\pi}{L}\right)^2(\tau_1-\tau_3)}e^{-\left(\frac{m\pi}{L}\right)^2(\tau_1-\tau_4)}e^{-\left(\frac{;\pi}{L}\right)^2(\tau_1-\tau_5)}e^{-\left(\frac{k\pi}{L}\right)^2(\tau_1-\tau_6)}e^{-\left(\frac{v\pi}{L}\right)^2(\tau_1-\tau_7)}e^{-\left(\frac{w\pi}{L}\right)^2(\tau_1-\tau_8)}\\
&\quad EF_{n_0}(\tau_3)F_{m_0}(\tau_4)F_{l_0}(\tau_5)F_{k_0}(\tau_6)F_{v_0}(\tau_7)F_{w_0}(\tau_8)\,d\tau_3d\tau_4d\tau_5d\tau_6d\tau_7d\tau_8
\end{aligned}
\tag{93}
$$

$$EI_{n_0}(\tau_1)I_{m_0}(\tau_1)I_{l_0}(\tau_1)I_{k_0}(\tau_2)I_{v_0}(\tau_2)I_{w_0}(\tau_2)I_{z_0}(\tau_2)I_{q_0}(\tau_2)$$

$$= \int_0^{\tau_2}\int_0^{\tau_2}\int_0^{\tau_2}\int_0^{\tau_2}\int_0^{\tau_2}\int_0^{\tau_1}\int_0^{\tau_1}\int_0^{\tau_1} e^{-\left(\frac{r\pi}{L}\right)^2(\tau_1-\tau_3)} e^{-\left(\frac{m\pi}{L}\right)^2(\tau_1-\tau_4)} e^{-\left(\frac{l\pi}{L}\right)^2(\tau_1-\tau_5)} e^{-\left(\frac{k\pi}{L}\right)^2(\tau_2-\tau_6)} e^{-\left(\frac{v\pi}{L}\right)^2(\tau_2-\tau_7)}$$

$$e^{-\left(\frac{w\pi}{L}\right)^2(\tau 2-\tau_8)} e^{-\left(\frac{z\pi}{L}\right)^2(\tau_2-\tau_9)} e^{-\left(\frac{q\pi}{L}\right)^2(\tau_2-\tau_{10})} EF_{n_0}(\tau_3)F_{m_0}(\tau_4)F_{l_0}(\tau_5)F_{k_0}(\tau_6)F_{v_0}(\tau_7)$$
$$F_{w_0}(\tau_8)F_{z_0}(\tau_9)F_{q_0}(\tau_{10})d\tau_3 d\tau_4 d\tau_5 d\tau_6 d\tau_7 d\tau_8 d\tau_9 d\tau_{10} \tag{94}$$

$$EI_{n_0}(\tau_1)I_{m_0}(\tau_1)I_{l_0}(\tau_1)I_{k_0}(\tau_1)I_{v_0}(\tau_1)I_{w_0}(\tau_2)I_{z_0}(\tau_2)I_{q_0}(\tau_2)I_{h_0}(\tau_2)I_{j_0}(\tau_2)$$

$$= \int_0^{\tau_2}\int_0^{\tau_2}\int_0^{\tau_2}\int_0^{\tau_2}\int_0^{\tau_2}\int_0^{\tau_1}\int_0^{\tau_1}\int_0^{\tau_1}\int_0^{\tau_1}\int_0^{\tau_1} e^{-\left(\frac{r\pi}{L}\right)^2(\tau_1-\tau_3)} e^{-\left(\frac{m\pi}{L}\right)^2(\tau_1-\tau_4)} e^{-\left(\frac{l\pi}{L}\right)^2(\tau_1-\tau_5)} e^{-\left(\frac{k\pi}{L}\right)^2(\tau_2-\tau_6)} e^{-\left(\frac{v\pi}{L}\right)^2(\tau_2-\tau_7)}$$

$$e^{-\left(\frac{w\pi}{L}\right)^2(\tau 2-\tau_8)} e^{-\left(\frac{z\pi}{L}\right)^2(\tau_2-\tau_9)} e^{-\left(\frac{q\pi}{L}\right)^2(\tau_2-\tau_{10})} e^{-\left(\frac{h\pi}{L}\right)^2(\tau_2-\tau_{11})} e^{-\left(\frac{j\pi}{L}\right)^2(\tau_2-\tau_{12})} EF_{n_0}(\tau_3)F_{m_0}(\tau_4)$$
$$F_{l_0}(\tau_5)F_{k_0}(\tau_6)F_{v_0}(\tau_7)F_{w_0}(\tau_8)F_{z_0}(\tau_9)F_{q_0}(\tau_{10})F_{h_0}(\tau_{11})F_{j_0}(\tau_{12})d\tau_3 d\tau_4 d\tau_5 d\tau_6 d\tau_7 d\tau_8 d\tau_9 d\tau_{10} d\tau_{11} d\tau_{12}, \tag{95}$$

Figures 21 and 22 illustrate the change of the mean and variance under the nonlinear scale parameter.

3.2. Using HPM Technique

Applying HPM on equation (38), one can get the following results w. r. t homotopy perturbation:

$$A(u) = R(u) + \varepsilon u^5 \tag{96}$$

$$R(u) = \frac{\partial u(t,x;\omega)}{\partial t} - \frac{\partial^2 u}{\partial x^2} \tag{97}$$

$$N(u) = \varepsilon u^5 \tag{98}$$

$$f(r) = \sigma \cdot n(x;\omega) \tag{99}$$

The homotopy function takes the following form:

$$H(s,p)=\left[R(s)-R(q_0)\right]+p\left[A(s)-f(r)\right]=0 \tag{100}$$

Or equivalently,

$$R(s)-R(q_0)+p\left[R(s_0)+\varepsilon s^5-\sigma\cdot n(x;\omega)\right]=0 \tag{101}$$

Where z_0 is an initial solution .The approximate solution can then be obtained using

$$s=s_0+ps_1+p^2s_2+p^3s_3+\cdots \tag{102}$$

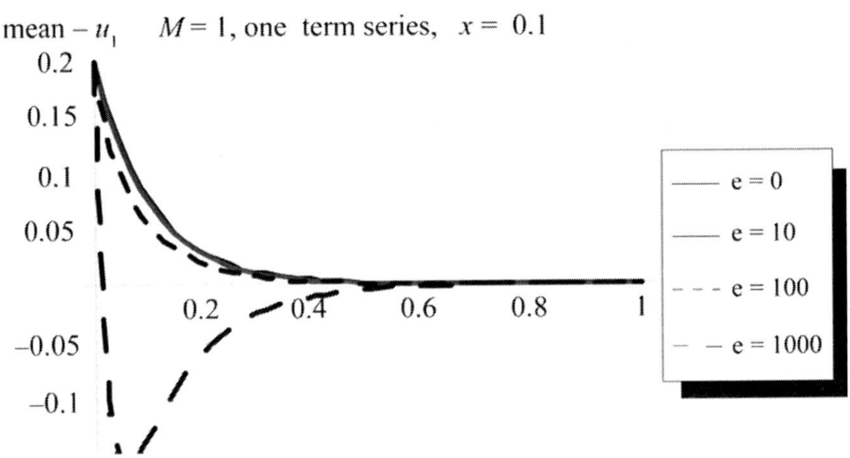

Figure 21. The change of the mean of the first order approximation $u_1$with time t at x = 0.1 For different ε values, (L = 1, M = 1, σ = 1, Φ(x) = x).

Now, setting p = 1, the approximation solution is obtained as:

$$u = \lim_{p \to 1} s = s_0 + s_1 + s_2 + s_3 + \cdots \tag{103}$$

Using equation (102) in equation (101) and equating the equal powers of p in both sides of the equation, one can get the following results:

i) $R(s_0) = R(q_0)$, in which one may consider the following simple solution:

$$q_0(t,0) = q_0(t,L) = 0,\ s_0 = q_0$$
$$\text{and } q_0(0,x) = \phi(x) \tag{104}$$

$$\textit{ii)}\ R(s_1) = \sigma \cdot n(x;\omega) - R(s_0) - \varepsilon s_0^5$$
$$s_1(t,0) = s_1(t,L) = 0, \text{ and } s_1(0,x) = 0 \tag{105}$$

$$\textit{iii)}\ R(s_2) = -5\varepsilon s_0^4 s_1$$
$$s_2(t,0) = s_2(t,L) = 0, \text{ and } s_2(0,x) = 0 \tag{106}$$

$$\textit{iv)}\ R(s_3) = -5\varepsilon(s_0^4 s_2 + 2 s_0^3 s_1^2)$$
$$s_3(t,0) = s_3(t,L) = 0, \text{ and } s_3(0,x) = 0 \tag{107}$$

$$vi)\ R(s_4) = -5\varepsilon(s_0^4 s_3 + 2s_0^2 s_1^3 + 4s_0^3 s_1 s_2)$$
$$s_4(t,0) = s_4(t,L) = 0, \text{ and } s_4(0,x) = 0 \tag{108}$$

As before we have many choices in guessing the initial approximation together with its initial conditions which greatly affects the consequent approximation. the choice s_0 is a design problem which can be taken as follows:

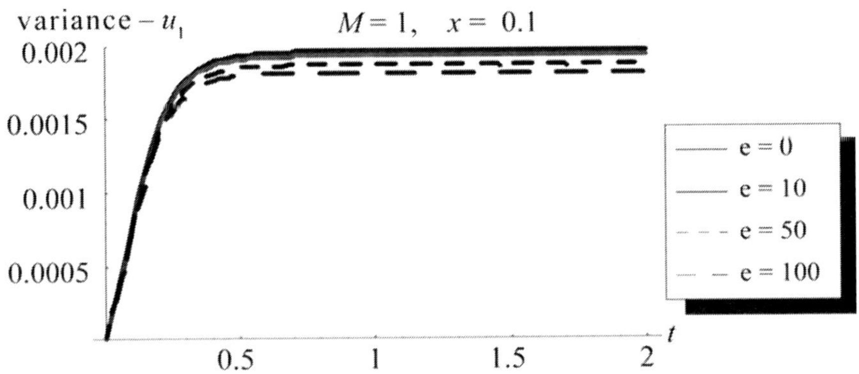

Figure 22. The change of the variance of the first order approximation u_1 with time t and space variable x at different ε values, (L = 1, M = 1, σ = 1, Φ(x) = x).

$$s_0(t,x) = \sum_{n=0}^{\infty} A_n e^{\beta_n t} \sin\frac{n\pi}{L}x$$
$$A_n = \int_0^L \phi(x) \sin\frac{n\pi}{L} x\, dx \tag{109}$$

First order approximation can be obtained using Eigenfunction expansion as follows:

$$u^{(1)}(t,x;s) = s_0 + s_1 = s_0 + \sum_{n=0}^{\infty} K_{n_1}(t)\sin\frac{n\pi}{L}x$$

where

$$K_{n_1}(t) = \int_0^t e^{\left(\frac{-n\pi}{L}\right)^2 (t-\tau)} M_{n_1}(\tau)\,d\tau$$

$$M_{n_1}(t) = \frac{2}{L}\int_0^L \left[\sigma \cdot n(x;s) - R(s_0) - \varepsilon s_0^5\right] dx \tag{110}$$

The ensemble average is:

$$\mu_{u^{(1)}}(t,x;s) = s_0 + \sum_{n=0}^{\infty} EK_{n_1}(t)\sin\frac{n\pi}{L}x$$

where

$$EK_{n_1}(t) = \int_0^t e^{\left(\frac{-n\pi}{L}\right)^2 (t-\tau)} EM_{n_1}(\tau)\,d\tau$$

$$EM_{n_1}(t) = \frac{-2}{L}\int_0^L \left[R(s_0) + \lambda s_0^5\right]\sin\frac{n\pi}{L}x\,dx \tag{111}$$

The covariance is obtained from the following final expression

$$Cov\left(u^{(1)}(t,x_1), u^{(1)}(t,x_2)\right)$$
$$= \frac{4\sigma^2}{L^2}\sum_{n=1}^{\infty}\sum_{m=1}^{\infty} \sin\frac{n\pi}{L}x_1 \sin\frac{m\pi}{L}x_2 \left(\int_0^L \sin\frac{n\pi}{L}x \sin\frac{m\pi}{L}x\,dx\right)$$
$$\left(\int_0^t\int_0^t e^{\left(\frac{-n\pi}{L}\right)^2 (t-\tau_1)} e^{\left(\frac{-m\pi}{L}\right)^2 (t-\tau_2)} d\tau_1 d\tau_2\right) \tag{112}$$

The variance can then be obtained from equation (112) by setting $x_1 = x_2 = x$.

Any higher order approximations can be obtained in a similar way. The following sample results using the same data in the cubic case.

3.3. Mean and Variance Comparisons of the Solution of Stochastic Quintic Nonlinear Diffusion Problem Using Pickard and HPM Methods with Different Orders and Corrections

Figures 23, 24 and 25 illustrate the change of the mean and variance for Picard and HPM approximations.

4. CONCLOUSION

Comparisons among results of the computations of mean illustrates that the results of three methods are very close from each other, but in variance the results of HPM and WHEP are semi-simillar while some results of Pickard are different in magnitude, may be due to the inability of computing high order approximations. It seems that the WHEP technique is more complex, in computations sense, than HPM which is more general since it can be applied to a non-perturbative problems as well as perturbative ones. The WHEP technique has the advantage of making different corrections for each order of approximation. The HPM seems easier in computations sense but it is very sensitive to initial guess solution. Symbolic computations with higher computations abilities will lead to better approximate solutions in the future.

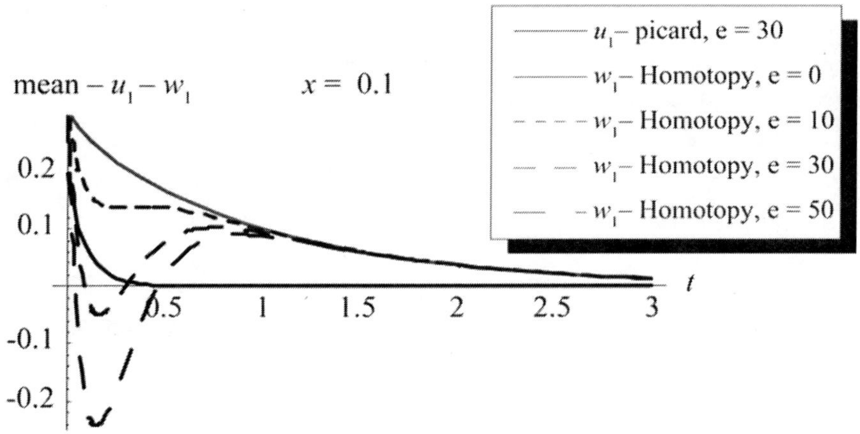

Figure 23. Mean comparison between Picard first order at ε = 30 and HPM first order at different ε values, (u^1, w^1) at x = 0.1, (L = 1, M = 1, σ = 1, Φ(x) = x).

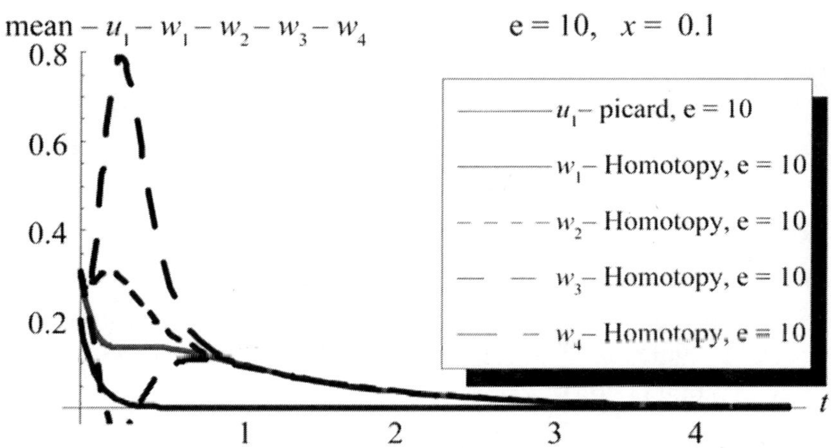

Figure 24. Mean comparison between Picard first order and HPM first, second, third and fourth order at ε = 10, (u^1, w^1, w^2, w^3, w^4) at x = 0.1, (L = 1, M = 1, σ = 1, Φ(x) = x).

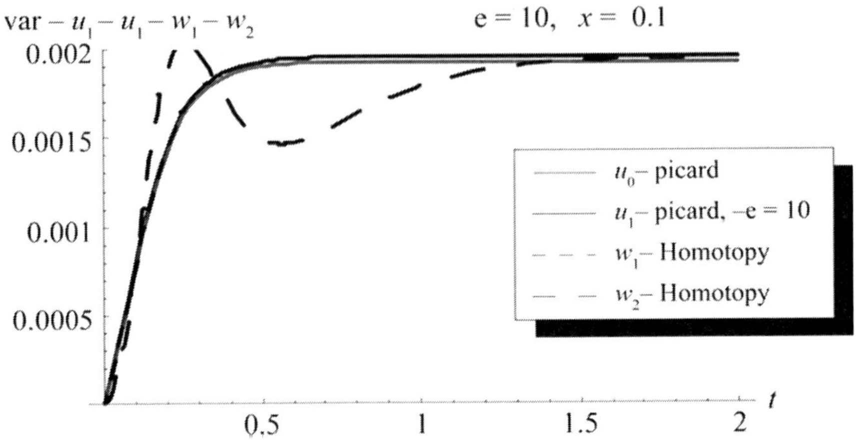

Figure 25. Variance comparison between Picard zero and first order, HPM first and second order, (u^0, u^1, w^1, w^2) at x = 0.1, ε = 10, (L = 1, M = 1, σ = 1, Φ(x) = x).

REFERENCES

1. Kampe de Feriet: Random solutions of partial differential equations, In: Proc. 3rd Berkeley Symposiumon Mathematical Statistics and Probability–1955, vol. III, pp. 199- 208 (1956).

2. Bharucha-Reid: A survey on the theory of random functions. The Institute of Mathematical Sciences: Matscience Report 31, India (1965).

3. Lo Dato, V.: Stochastic processes in heat and mass transport, C. In: Bharucha-Reid (ed.) Probabilistic Methods in Applied Mathematics, vol. 3, pp. 183-212. Academic, New York (1973).

4. Becus, A. G.: Random generalized solutions to the heat equations. J. Math. Anal. Appl. 60, 93-102 (1977).

5. Marcus, R.: Parabolic Ito equation with monotone nonlinearities. J. Funct. Anal. 29, 257-286 (1978).

6. Manthey, R.: Weak convergence of solutions of the heat equation with Gaussian noise. Math. Nachr. 123, 157-168 (1985).

7. Manthey, R.: Existence and uniqueness of a solution of a reaction-diffusion with polynomial nonlinearity and with noise disturbance. Math. Nachr. 125, 121-133 (1986).

8. Jetschke, G.: II. Most probable states of a nonlinear Brownian bridge. Forschungsergebnisse (Jena) N/86/20 (1986).

9. Jetschke, G.: III. Tunneling in a bistable infinitedimensional potential. Forschungsergebnisse (Jena) N/86/40 (1986).

10. El-Tawil, M.: Nonhomogeneous boundary value problems. J. Math. Anal. Appl. 200, 53-65 (1996).

11. Uemura, H.: Construction of the solution of 1-dim heat equation with white noise potential and its asymptotic behaviour. Stoch. Anal. Appl. 14(4), 487-506 (1996).

12. El-Tawil, M.: The application ofWHEP technique on partial differential equations. Int. J. Differ. Equ. Appl. 7(3), 325-337 (2003).

13. El-Tawil, M.: The homotopy Wiener-Hermite expansion and perturbation technique (WHEP). In: Transactions on Computational Science I. LNCS, vol. 4750, pp. 159-180. Springer, New York (2008).

14. Crow, S., Canavan, G.: Relationship between a WienerHermite expansion and an energy cascade. J. Fluid Mech. 41(2), 387-403 (1970).

15. El-Tawil M. and Noha A. El-Molla, The approximate solution of a nonlinear diffusion equation using some techniques, a comparison study, International Journal of Nonlinear Sciences and numerical Simulation, 10(3), 687-698, 2009.

16. El-Tawil M. and Noha A. El-Molla, Solving nonlinear diffusion equations without stochastic homogeneity using the homotopy perturbation method, J. of applied mathematics, pp. 281-299, 2009.

17. Farlow, S. J.: Partial Differential Equations for Scientists and Engineers. Wiley, New York (1982).

18. He, J. H.: Homotopy perturbation technique. Comput. Methods Appl. Mech. Eng. 178, 257-292(1999).

19. He, J. H.: A coupling method of a homotopy technique and a perturbation technique for nonlinear problems. Int. J. Nonlinear Mech. 35, 37-43 (2000).

20. He, J. H.: Homotopy perturbation method: a new nonlinear analytical technique. Appl.Math. Comput. 135, 73-79 (2003).

21. He, J. H.: The homotopy perturbation method for nonlinear oscillators with discontinuities. Appl.Math. Comput. 151, 287-292 (2004).

22. El-Tawil M., International Journal of Differential Equations and its Applications, Vol. 7, No. 3, pp 325-337, 2003.

23. El-Tawil M., The Homotopy Wiener-Hermite expansion and perturbation technique (WHEP), Transactions on Computational Science (Springer), accepted.

24. Farlow S. J., Partial differential equations for scientists and engineers, Wiley, N. Y., 1982.

25. Crow S. and Canavan G., Relationship between a Wiener-Hermite expansion and an energy cascade, J. of fluid mechanics, 41(2), pp. 387-403 (1970).

26. Saffman P., Application of Wiener-Hermite expansion to the diffusion of a passive scalar in a homogeneous turbulent flow, Physics of fluids, 12(9), pp. 1786-1798(1969).

27. Kahan W. and Siegel A., Cameron-Martin-Wiener method in turbulence and in Burger's model: General formulae and application to late decay, J. of fluid mechanics, 41(3), pp. 593-618 (1970).

28. Chorin and Alexandre J., Gaussian fields and random flow, J. of fluid of mechanics, 63(1), pp. 21-32(1974).

29. Eftimiu and Cornel, First-order Wiener-Hermite expansion in the electromagnetic scattering by conducting rough surfaces, Radio science, 23(5), pp. 769-779(1988).

30. Jahedi A. and Ahmadi G., Application of Wiener-Hermite expansion to non-stationary random vibration of a Duffing oscillator, J. of applied mechanics, Transactions ASME, 50(2), pp. 436-442(1983).

31. Tamura Y. and Nakayama J., A formula on the Hermite expansion and its aoolication to a random boundary value problem, IEICE Transactions on electronics, E86-C(8), pp. 1743-1748 (2003).

32. Kayanuma Y. and Noba K., Wiener-Hermite expansion formalism for the stochastic model of a driven quantum system, Chemical physics, 268(1-3), pp. 177-188(2001).

33. Gawad E. and El-Tawil M., General stochastic oscillatory systems, Applied Mathematical Modelling, 17(6), pp. 329-335(1993).

34. Imamura T. Meecham W. and Siegel A., Symbolic calculus of the Wiener process and Wiener-Hermite functionals, J. math. Phys., 6(5), pp. 695-706(1983).

6

CHAPTER

Stochastic Approximation Method for Fixed Point Problems

Ya. I. Alber[1], C. E. Chidume[2], Jinlu Li[3]

[1]*Department of Mathematics, The Technion-Israel Institute of Technology, Haifa, Israel*

[2]*The Abdus Salam International Centre for Theoretical Physics, Trieste, Italy*

[3]*Department of Mathematical Sciences, Shawnee State University, Portsmouth, USA*

ABSTRACT

We study iterative processes of stochastic approximation for finding fixed points of weakly contractive and nonexpansive operators in Hilbert spaces under the condition that operators are given with random errors. We prove mean square convergence and convergence almost sure (a.s.) of iterative approximations and establish both asymptotic and nonasymptotic estimates of the convergence rate in degenerate and non-degenerate cases. Previously the stochastic approximation algorithms were studied mainly for optimization problems.

Keywords: Hilbert Spaces; Stochastic Approximation Algorithm; Weakly Contractive Operators; Nonexpansive Operators; Fixed Points; Convergence in Mean Square; Convergence Almost Sure (a.s.); Nonasymptotic Estimates of Convergence Rate

1. INTRODUCTION

In this paper the following problem is solved: To find a fixed point x^* of the operator $T:G \to H$, in other words, to find a solution $x^* \in G$ of the equation

$$x = Tx, \tag{1.1}$$

where T is a Lipschitz continuous mapping, H is a Hilbert space, $G \subseteq H$ is a closed convex subset. We suppose that x^* exists, i.e., the fixed point set N of T is nonempty. Note in different particular cases of the Equation (1.1), for example, when $T:G \to G$, the solution existence and solution uniqueness can be proved under some additional assumptions.

We separately consider two classes of mappings T: the class of weakly contractive maps and more general class of nonexpansive ones. Let us recall their definitions.

Definition 1.1. A mapping $T:G \to H$ is said to be weakly contractive of class $C_{\phi(t)}$ on a closed convex subset $G \subset H$ if there exists a continuous and increasing function $\phi(t)$ defined on IR^+ such that ϕ is positive on $IR^+/\{0\}, \phi(0)=0, \lim_{t\to+\infty}\phi(t)=+\infty,$ and for all

$$x, y \in G, \|Tx - Ty\| \le \|x - y\| - \phi(\|x - y\|). \tag{1.2}$$

Remark 1.2. It follows from (1.2) that $\psi(t) \le t$ and in real problems an argument t of the function $\psi(t)$ doesn't necessary approaches to ∞ obeying the condition $T:G \to H$ (see the example in Remark 3.4).

Definition 1.3. A mapping $T: G \to H$ is said to be nonexpansive on the closed convex subset $G \subset H$ if for all $x, y \in G$

$$\|Tx - Ty\| \le \|x - y\|.$$

It is obvious that the class of weakly contractive maps is contained in the class of nonexpansive maps because the right-hand side of (1.2) is estimated as

$$0 \le \|x - y\| - \phi(\|x - y\|) \le \|x - y\|, \tag{1.3}$$

and it contains the class of strongly contractive maps because $\phi(t) = (1-q)t$ with $0 < q < 1$ gives us

$$\|Tx - Ty\| \le q\|x - y\|. \tag{1.4}$$

We study the following algorithm of stochastic approximation:

$$x_{n+1} = Pr_G(x_n - \alpha_n S_n x_n), n = 1, 2, \cdots, x_1 \in G, \tag{1.5}$$

where Pr_G is the metric projection operator from H onto G and deterministic step-parameters α_n satisfy the standard conditions:

$$\sum_{n=1}^{\infty} \alpha_n = \infty \text{ and } \sum_{n=1}^{\infty} \alpha_n^2 < \infty. \tag{1.6}$$

The factor $S_n x_n$ in (1.5) is an infinite-dimensional vector of random observations of the clearance operator $F = I - T$ at random points $x_n \in G$ given for all $n \geq 1$ on the same probability space (Ω, A, P). We set

$$S_n x_n = \zeta_n + \xi_n, \tag{1.7}$$

where $\zeta_n = x_n - Tx_n$ and ξ_n is a sequence of independent random vectors with the conditions

$$E[\xi_n] = 0 \text{ and } E\left[\|\xi_n\|^2\right] \leq C_1, 0 < C_1 < \infty. \tag{1.8}$$

Here E is a symbol of the mathematical expectation. In order to calculate conditional mathematical expectations of different random variables we define the σ - subalgebra $A_n := \sigma(x_1, x_2, \cdots, x_n)$ on (Ω, A, P). And then $E[\xi_n / A_n]$ means A_n-measurable function with the following property: for any $B \in A_n$

$$\int_B \xi_n \mathrm{d}P(\omega) = \int_B E[\xi_n / A_n] \mathrm{d}P(\omega).$$

We also assume in the sequel that ζ_n is A_n-measurable for all $n \geq 1$.

Let us recall the mean square convergence and almost sure (a.s.) convergence.

We say that the sequence $\{\xi_n\}$ of random variables $\xi_n(\omega)$ converges in mean square to ξ if ξ exists and

$$\lim_{n\to\infty} E\left[\left\|\xi_n - \xi\right\|^2\right] = 0.$$

The sequence ξ_n converges to ξ almost surely or with probability 1 if

$$P\left(\lim_{n\to\infty} \xi_n(\omega) = \xi(\omega)\right) = 1.$$

Almost sure convergence and convergence in mean square imply convergence in the sense of probability: The sequence $\{\xi_n\}$ of random variables $\xi_n(\omega)$ converges in the sense of probability to $\xi(\omega)$ if for all $\varepsilon > 0$

$$\lim_{n\to\infty} P\left(\left\|\xi_n(\omega) - \xi(\omega)\right\| \geq \varepsilon\right) = 0.$$

So, we consider iterative processes of stochastic approximation in the form (1.5) for finding fixed points of weakly contractive (Definition 1.1) and nonexpansive (Definition 1.3) mappings in Hilbert spaces under the conditions (1.8). We prove mean square convergence and convergence almost sure of iterative approximations and establish both asymptotic and nonasymptotic estimates of the convergence rate. Perhaps, we present here the first results of this sort for fixed point problems. Formerly the stochastic approximation methods were studied mainly to find minimal and maximal points in optimization problems (see, for example, [1-6] and references within).

2. AUXILIARY RECURRENT INEQUALITIES

Lemma 2.1. [3,4] Let $\{\mu_k\}, \{\rho_k\}$ and $\{\gamma_k\}$ be sequences of nonnegative real numbers satisfying the recurrent inequality.

$$\mu_{k+1} \leq (1+\rho_k)\mu_k + \gamma_k, k = 1,2,\cdots \tag{2.1}$$

Assume that $\sum_{k=1}^{\infty}\rho_k < \infty$ and $\sum_{k=1}^{\infty}\gamma_k < \infty.$ Then $\{\mu_k\}$

is bounded and converges to some limit.

Lemma 2.2. [3,4] Let $\{\mu_k\}, \{\zeta_k\}, \{\alpha_k\}$ and $\{\gamma_k\}$ be sequences of nonnegative real numbers satisfying the recurrent inequality.

$$\mu_{k+1} \leq (1+\rho_k)\mu_k - \alpha_k\psi(\mu_k) + \gamma_k, k = 1,2,\cdots, \tag{2.2}$$

where $\sum_{k=1}^{\infty}\alpha_k = \infty, \sum_{k=1}^{\infty}\rho_k < \infty$ and either $\sum_{k=1}^{\infty}\gamma_k < \infty$ or

$$\lim_{k\to\infty}\frac{\gamma_k}{\alpha_k} = 0. \tag{2.3}$$

Assume that $\psi(t)$ is continuous and increasing function defined on IR^1 such that ψ is positive on $IR^1/\{0\}$, $\psi(0)=0.$ Then $\lim_{k\to\infty}\mu_k = 0.$ There exists an infinite subsequence $\{k_l\}, l = 1,2,\cdots,$ such that

$$\mu_{k_l} \leq C_0\psi^{-1}\left(\frac{1}{\sum_{k=1}^{k_l}\alpha_k} + \frac{\gamma_{k_l}}{\alpha_{k_l}}\right),$$

where $C_0 \geq \prod_{k=1}^{\infty}(1+\rho_k).$

In the following two lemmas we want to present nonasymptotic estimates for the whole sequence $\mu_k, k \geq 1$. For this the stronger requirements are made of parameters α_k and function $\psi(t)$ in the recurrent inequality.

Suppose that $\alpha(t)$ such that $\alpha(k)=\alpha_k$, $F(t)$ and $\Phi(t)$ are antiderivatives from $\alpha(t)$ and $\frac{1}{\psi(t)}$, respectively, with arbitrary constants C (without loss of generality, one can put $C=0$), i.e.

$$F(t)=\int \alpha(t)\mathrm{d}t, \Phi(t)=\int \frac{\mathrm{d}t}{\psi(t)}.$$

Observe that $F(t)$ has the following properties:

i) $F'(t)=\alpha(t)$;

ii) $F(t)$ is strictly increasing on $[1,\infty)$ and $F(t)\to\infty$ as $t\to\infty$;

iii) The function $g(t)=\frac{\alpha(t)}{F(t)}=\frac{F'(t)}{F(t)}$ is decreasing;

iv) $G(t)=\ln F(t)\to\infty$ as $t\to\infty$.

Introduce the following denotations:

1) $\psi^{-1}(z)$ and $\Phi^{-1}(z)$ are the inverse functions to $\psi(t)$ and $\Phi(t)$, respectively;

2) $v(s)=\psi^{-1}\left[\frac{c_0\gamma(s-1)}{\alpha(s-1)}\right]+\gamma(s-1), s\geq 2, c_0>1$ is a fixed control parameter;

3) $u(s,C)=\Phi^{-1}\left[\Phi(C)-a\left(F(s)-F(2)\right)\right]$,

$C \geq 0, a = \frac{c_0 - 1}{c_0} < 1;$

4) $w(s, v(r)) = \Phi^{-1}\left[\Phi(v(r)) - a(F(s) - F(2))\right],$

$2 \leq r < \infty;$

5) $Q(\bar{c}) = \Phi^{-1}\left[\Phi(\bar{c}) - a(F(2) - F(1))\right],$ where

$\bar{c} > 0$ is an arbitrary fixed number;

We present now the based condition (P): The graphs of the scalar functions $v(s)$ and $w(s, v(r))$ with any fixed $r \in [2, \infty)$ are intersected on the interval $[2, \infty)$ not more than at two points s_1 and s_2 (we do not consider contact points as intersection ones excepting $s = 2$ if any).

For example, the graphs of the functions $v(s)$ and

$w(s, v(r))$ calculated for $\alpha(s) = \frac{b}{s^{\kappa}}, b > 0, 0 < \kappa \leq 1,$

and $\psi(t) = t^{\nu}, \nu \geq 1,$ satisfy the condition (P).

Lemma 2.3. [3,4] Assume that 1) the property (P) is carried out for the function $w(s, v(2)) = u(s, v(2))$ and

$v(s)$; 2) $u(s, v(2)) \geq v(s)$ as $s \to \infty$; 3) the control parameter c_0 is chosen such that

$$u(s, v(2)) \geq v(s) \text{ as } s \to 2. \tag{2.4}$$

Then for the sequence $\{\mu_k\}$ generated by the inequality

$$\mu_{k+1} \leq \mu_k - \alpha_k \psi(\mu_k) + \gamma_k, k = 1, 2, \cdots, \tag{2.5}$$

it follows: $\lim_{k\to\infty} \mu_k = 0$ and for all $k \geq 1$

$$\mu_k \leq u(k, C), C = \max\{Q(\mu_1), v(2)\}. \tag{2.6}$$

Lemma 2.4. [3,4] Assume that 1) the property (P) is carried out for all the function $w(s, v(2r))$ and $v(s)$; 2) $u(s, v(2)) \leq v(s)$ as $s \to \infty$. Then for the sequence $\{\mu_k\}$ generated by the inequality (2.5) $\lim_{k\to\infty}\mu_k = 0$. In additiona) if $Q(\mu_1) \leq v(2)$ and the control parameter c_0 is chosen such that $u(s, v(2)) \leq v(s)$ as $s \to 2$, then for all $k \geq 2$

$$\mu_k \leq v(k); \tag{2.7}$$

b) in all remaining cases

$$\mu_k \leq u(k, C), C = \max\{Q(\mu_1), v(2)\},$$
$$1 \leq k \leq \bar{s}, \tag{2.8}$$

$$\mu_k \leq v(k), k > \bar{s}, \tag{2.9}$$

where $\bar{s}$ is a unique root of the equation

$$u(s, C) = v(s) \tag{2.10}$$

on the interval $[2, \infty)$.

The following lemmas deal with another sort of recurrent inequalities:

Lemma 2.5. [7,8] Let $\{\mu_k\}, \{\alpha_k\}, \{\beta_k\}$ and $\{\gamma_k\}$ be sequences of non-negative real numbers satisfying the recurrence inequality.

$$\mu_{k+1} \leq \mu_k - \alpha_k \beta_k + \gamma_k, k = 1, 2, \cdots. \tag{2.11}$$

Assume that

$$\sum_{k=1}^{\infty} \alpha_k = \infty \text{ and } \sum_{k=1}^{\infty} \gamma_k < \infty.$$

Then:

i) There exists an infinite subsequence $\{\beta_{\ell_k}\} \subset \{\beta_k\}$ such that

$$\beta_{\ell_k} \leq \frac{1}{\sum_{j=1}^{\ell_k} \alpha_j}, \tag{2.12}$$

and, consequently, $\lim_{k\to\infty} \beta_{\ell_k} = 0;$

ii) if $\lim_{k\to\infty} \alpha_k = 0$ and there exists $\kappa > 0$ such that

$$\left| \beta_{k+1} - \beta_k \right| \leq \kappa \alpha_k \tag{2.13}$$

for all $k \geq 1$, then $\lim_{k\to\infty} \beta_k = 0.$

Lemma 2.6. [7,8] Let $\{\mu_k\},\{\alpha_k\},\{\beta_k\}$ and $\{\gamma_k\}$ be sequences of non-negative real numbers satisfying the recurrence inequality (2.11). Assume that $\sum_{k=1}^{\infty}\alpha_k=\infty$

and (2.3) is satisfied. Then there exists an infinite subsequence $\{\beta_{t_k}\}\subset\{\beta_k\}$ such that $\lim_{k\to\infty}\beta_{t_k}=0.$

3. MEAN SQUARE CONVERGENCE OF STOCHASTIC APPROXIMATIONS

Theorem 3.1. Assume that $T:G\to H$ is a weakly contractive mapping of the class $C_{\phi(\tau)}$, $\psi_1(\tau)=\tau\phi(\tau)$ is a convex function with respect to $t=\tau^2$ and $E\left[\|x_1-x^*\|^2\right]<\infty.$ Then the sequence $\{x_n\}$ generated by (1.5)-(1.7) converges in mean square to a unique fixed point x^* of T. There exists an infinite subsequence $\{n_l\}, l=1,2,\cdots,$ such that

$$E\left[\|x_{n_l}-x^*\|^2\right]\le C_0\psi^{-1}\left(\frac{1}{2\sum_{n=1}^{n_l}\alpha_n}+C_1\alpha_{n_l}\right), \tag{3.1}$$

where $\psi(\tau^2)=\psi_1(\tau)$ and some positive constant C_0 satisfies the inequality

$$\prod_{n=1}^{\infty}\left(1+8\alpha_n^2\right)\le C_0. \tag{3.2}$$

Remark 3.2. $1<C_0<\infty$ exists in view of the second condition in (1.6).

Proof. First of all, we note that the method (1.5)-(1.7) guarantees inclusion $x_n\in G$ for all $n\ge 1$. Since the metric projection operator Pr_G is nonexpansive in a Hilbert space and $x^*\in G$ exists, we can write

$$\begin{aligned}
&\left\|x_{n+1}-x^*\right\|^2\\
&=\left\|Pr_G\left(x_n-\alpha_n\left(x_n-Tx_n+\xi_n\right)\right)-Pr_G x^*\right\|^2\\
&\le\left\|x_n-x^*\right\|^2-2\alpha_n\left(\zeta_n+\xi_n,x_{n+1}-x^*\right)\\
&=\left\|x_n-x^*\right\|^2-2\alpha_n\left(\zeta_n,x_n-x^*\right)\\
&-2\alpha_n\left(\xi_n,x_n-x^*\right)+\left\|x_{n+1}-x_n\right\|^2.
\end{aligned} \tag{3.3}$$

Let us evaluate the first scalar product in (3.3). We have

$$\begin{aligned}
&\left(x_n-Tx_n,x_n-x^*\right)\\
&=\left(x_n-x^*-Tx_n+Tx^*,x_n-x^*\right)\\
&=\left\|x_n-x^*\right\|^2-\left(Tx_n-Tx^*,x_n-x^*\right)\\
&\ge\left\|x_n-x^*\right\|^2-\left\|Tx_n-Tx^*\right\|\left\|x_n-x^*\right\|\\
&\ge\left\|x_n-x^*\right\|^2-\left(\left\|x_n-x^*\right\|-\phi\left(\left\|x_n-x^*\right\|\right)\right)\left\|x_n-x^*\right\|\\
&\ge\phi\left(\left\|x_n-x^*\right\|\right)\left\|x_n-x^*\right\|=\psi_1\left(\left\|x_n-x^*\right\|\right)
\end{aligned} \tag{3.4}$$

We remember that $\psi\left(\tau^2\right)=\psi_1(\tau)$. Then the inequalities (3.3) and (3.4) yield

$$\begin{aligned}
&\left\|x_{n+1}-x^*\right\|^2\le\left\|x_n-x^*\right\|^2-2\alpha_n\psi\left(\left\|x_n-x^*\right\|^2\right)\\
&-2\alpha_n\left(\xi_n,x_n-x^*\right)+\alpha_n^2\left\|\zeta_n+\xi_n\right\|^2.
\end{aligned} \tag{3.5}$$

Applying the conditional expectation with respect to A_n to the both sides of (3.5) we obtain

$$\begin{aligned}
&E\left[\left\|x_{n+1}-x^*\right\|^2/A_n\right]\\
&\le\left\|x_n-x^*\right\|^2-2\alpha_n\psi\left(\left\|x_n-x^*\right\|^2\right)\\
&-2\alpha_n E\left[\left(\xi_n,x_n-x^*\right)/A_n\right]+\alpha_n^2 E\left[\left\|\zeta_n+\xi_n\right\|^2/A_n\right].
\end{aligned} \tag{3.6}$$

It is easy to see that

$$E\left[\left\|\zeta_n+\xi_n^2\right\|/A_n\right]$$
$$\leq 2E\left[\left\|\zeta_n\right\|^2/A_n\right]+2E\left[\left\|\xi_n\right\|^2/A_n\right]$$
$$=2\left\|\zeta_n\right\|^2+2E\left[\left\|\xi_n\right\|^2/A_n\right]. \tag{3.7}$$

Since T is weakly contractive and therefore nonexpansive, one gets

$$\left\|x_n-Tx_n\right\|^2\leq 4\left\|x_n-x^*\right\|^2.$$

Taking into account (3.7), the inequality (3.9) is estimated as follows:

$$E\left[\left\|x_{n+1}-x^*\right\|^2/A_n\right]$$
$$\leq\left(1+8\alpha_n^2\right)\left\|x_n-x^*\right\|^2-2\alpha_n\psi\left(\left\|x_n-x^*\right\|^2\right)$$
$$-2\alpha_nE\left[\left(\xi_n,x_n-x^*\right)/A_n\right]+2\alpha_n^2E\left[\left\|\xi_n\right\|^2/A_n\right]. \tag{3.8}$$

Now the unconditional expectation implies

$$E\left[\left\|x_{n+1}-x^*\right\|^2\right]$$
$$\leq\left(1+8\alpha_n^2\right)E\left[\left\|x_n-x^*\right\|^2\right]-2\alpha_nE\left[\psi\left(\left\|x_n-x^*\right\|^2\right)\right]$$
$$-2\alpha_nE\left[\left(\xi_n,x_n-x^*\right)\right]+2\alpha_n^2E\left[\left\|\xi_n\right\|^2\right] \tag{3.9}$$

Next we need the Jensen inequality for a convex function $\psi\left(\left\|x_n-x^*\right\|^2\right)$:

$$E\left[\psi\left(\left\|x_n-x^*\right\|^2\right)\right]\leq\psi\left(E\left[\left\|x_n-x^*\right\|^2\right]\right)$$

(see [9,10]). This allows us to rewrite (3.9) in the form

$$E\left[\left\|x_{n+1}-x^*\right\|^2\right] \le \left(1+8\alpha_n^2\right)E\left[\left\|x_n-x^*\right\|^2\right] - 2\alpha_n\psi\left(E\left[\left\|x_n-x^*\right\|^2\right]\right)+2\alpha_n^2E\left[\left\|\xi_n\right\|^2\right] \tag{3.10}$$

because of

$$E\left[\left(\xi_n, x_n-x^*\right)\right]=0.$$

Denoting $\lambda_n = E\left[\left\|x_n-x^*\right\|^2\right]$ we have

$$\lambda_{n+1} \le \left(1+8\alpha_n^2\right)\lambda_n - 2\alpha_n\psi\left(\lambda_n\right)+2C_1\alpha_n^2, \tag{3.11}$$

where in view of Definition 1.1, $\psi(\lambda)$ is a continuous and increasing function with $\psi(0)=0$. Due to (6), from Lemma 2.2 it follows

$$\lim_{n\to\infty} E\left[\left\|x_n-x^*\right\|^2\right]=0$$

and the estimate (3.1) holds too. The theorem is proved. □

Remark 3.3. If a fixed point of weakly contractive mapping $T:G\to H$ exists, then it is unique [11].

Remark 3.4. The following example was presented in [11]: Let $Tx=\sin x$, $G=[0,1]$ and $T:G\to G$. It has been shown in [11] that

$$|\sin x-\sin y|\le |x-y|-\frac{1}{8}|x-y|^3$$

for all $0\le x\le y\le 1$. Then

$$\phi(\tau)=\frac{1}{8}\tau^3,\psi_1(\tau)=\frac{1}{8}\tau^4 \text{ and } \psi(t)=\frac{1}{8}t^2.$$

Definition 3.5. Let a nonexpansive mapping

$T:G\to H$ have a unique fixed point x^*. T is said to be weakly sub-contractive on the closed convex subset $G\subset H$ if there exists continuous and increasing function $\psi(t)$ defined on IR^+ such that ψ is positive on

$IR^1/\{0\}$, $\psi(0)=0$, $\lim_{t\to+\infty}\phi(t)=+\infty$, such that for all

$x\in G$.

$$\langle x-Tx,x-x^*\rangle\ge\psi\left(\|x-x^*\|^2\right). \tag{3.12}$$

Theorem 3.6. Assume that a mapping $T:G\to H$ is weakly sub-contractive and the function $\psi(t)$ in (3.12) is convex on IR^+ Then the results of Theorem 3.1 holds for the sequence $\{x_n\}$ generated by (1.5)-(1.7).

The second inequality in (1.6) can be omitted if we assume not less than linear growth of $\psi(\lambda)$ "on infinity" and put $\alpha_n\to 0$ as $n\to\infty$

Theorem 3.7. Assume that a mapping $T: G \to H$ is weakly sub-contractive and the function $\psi(t)$ in (3.12) is convex on IR^+ Suppose that instead of (1.6) the conditions

$$\lim_{n\to\infty} \alpha_n = 0 \text{ and } \sum_{n=1}^{\infty} \alpha_n = \infty. \tag{3.13}$$

hold. In addition, let $\alpha_n \le \bar{\alpha} < 0.5$ and

$$\lim_{\lambda\to\infty} \frac{\psi(\lambda)}{\lambda} \ge 4\bar{\alpha}. \tag{3.14}$$

Then the sequence $\{x_n\}$ generated by (1.5), (1.7) and (3.13) converges in mean square to x^*. There exists an infinite subsequence $\{n_l\}, l = 1, 2, \cdots,$ such that

$$E\left[\left\|x_{n_l} - x^*\right\|^2\right] \le \psi^{-1}\left(\frac{1}{2\sum_{n=1}^{n_l} \alpha_n} + \frac{C_3 \alpha_{n_l}}{2}\right).$$

where

$$\begin{aligned} C_3 &= 2C_1 + 8\max\left\{\left\|x^* - x_1\right\|^2, C_2\right\}, \\ C_2 &= \lambda_* + 2C_1\bar{\alpha}^2 + 8\bar{\alpha}^2\lambda_*, \\ \lambda_* &= \psi^{-1}\left(C_1\bar{\alpha} + 4\bar{\alpha}\lambda_*\right). \end{aligned} \tag{3.15}$$

Proof. Consider the inequality (11) in the form

$$\lambda_{n+1} \le \lambda_n - 2\alpha_n \psi(\lambda_n) + 2C_1\alpha_n^2 + 8\alpha_n^2 \lambda_n, \tag{3.16}$$

where $\lambda_n = E\left[\left\|x_n - x^*\right\|^2\right].$ Observe that it is derived by making use of (3.4) and the nonexpansivity property of $T.$ We shall show that λ_n are bounded for all $n \in I = \{1,2,\cdots\}.$ Indeed, since $\psi(\lambda)$ is a convex increasing continuous function, we conclude that

$\psi_1(\lambda) = \frac{\psi(\lambda)}{\lambda}$ is nondecreasing and since (3.14) holdsthe inequality $\psi(\lambda) \le C_1\bar{\alpha} + 4\bar{\alpha}\lambda$ has a solution $\lambda \le \lambda_*,$ where λ_* is the unique root of the scalar equ0 ation $\psi(\lambda) = C_1\bar{\alpha} + 4\bar{\alpha}\lambda.$ Together with this, (3.4) and (3.14) are co-ordinated by the parameter $\bar{\alpha} < 0.5.$

Only one alternative can happen for each $n \in I:$ either

$$H_1 := -2\alpha_n\psi(\lambda_n) + 2C_1\alpha_n^2 + 8\alpha_n^2\lambda_n \ge 0$$

or

$$H_2 := -2\alpha_n\psi(\lambda_n) + 2C_1\alpha_n^2 + 8\alpha_n^2\lambda_n < 0.$$

Denote $I_1 = \{n \in I | H_1 \text{ is true}\}$ and $I_2 = \{n \in I | H_2 \text{ is true}\}$. It is clear that $I_1 \cup I_2 = I.$ From the hypothesis $H_1,$ it arises

$\psi(\lambda_n) \le C_1\bar{\alpha} + 4\bar{\alpha}\lambda_n,$ and then $\lambda_n \le \lambda_*$ for all $n \in I_1.$ From the hypothesis $H_2,$ we have: $\lambda_{n+1} < \lambda_n$ for all $n \in I_2.$ Consider all the possible cases:

1) $I_2 = \varnothing.$ Then $\lambda_n \le \lambda_*$ for all $n \in I.$

2) $I_1 = \varnothing$. Then $\lambda_n \le \lambda_1$ for all $n \in I$.

3) Let $I_1 = \{1, 2, \cdots, N_0\}$ and $I_2 = \{N_0 + 1, N_0 + 2, \cdots\}$. Then $\lambda_n \le \lambda_*$ for $n = 1, 2, \cdots, N_0$. By (3.16), $\lambda_{N_0+1} \le C_2$. It is obvious that $\lambda_n \le C_2$ for $N_0 + 2, N_0 + 3, \cdots$. Therefore, $\lambda_n \le C_2$ for all $n \in I$.

4) Let $I_2 = \{1, 2, \cdots, N_0\}$ and $I_1 = \{N_0 + 1, N_0 + 2, \cdots\}$. Then $\lambda_n \le \lambda_1$ for $n = 1, 2, \cdots, N_0$. and $\lambda_n \le \lambda_*$ for $N_0 + 1, N_0 + 2, \cdots$ Thus, $\lambda_n \le \lambda_*$ for all $n \in I$.

5) Let I_1 and I_2 be unbounded sets. Consider an arbitrary interval

$$I_s = [n_s + 1, n_{s+1} - 1] \subset I_2,$$

where $n_s, n_{s+1} \in I_1, s = 1, 3, 5, \cdots$. It is easy to be sure that $\lambda_{n_s} \le \lambda_*$ and $\lambda_n \le C_2$ for all $n \in I_s$.

6) The other situations of bounded and unbounded sets I_1 and I_2 are covered by the items 1)-5). Consequently, we have the final result: $\lambda_n \le \max\{\lambda_1, C_2\}$ for all $n \in I$.

Thus, we obtain the inequality

$$\lambda_{n+1} \le \lambda_n - 2\alpha_n \psi(\lambda_n) + C_3 \alpha_n^2, \tag{3.17}$$

where C_3 is defined by (3.15). Now Lemma 2.2 with the condition (2.3) implies the result. □

Remark 3.8. For a linear function $\psi(\tau) = c\tau$ which is convex and concave at the same time we suppose $2 \ge c > 4\bar{\alpha}$.

Remark 3.9. If G is bounded or more generally $\{x_n\}$ is bounded, then the inequality (3.17) (with some different constant C_3) immediately follows from (3.16).

4. ESTIMATES OF THE MEAN SQUARE CONVERGENCE RATE

Using Lemmas 2.3 and 2.4 we are able to give two general theorems on the nonasymptotic estimates of the mean square convergence rate for sequence $\{x_n\}$ generated by the stochastic approximation algorithm (1.5)- (1.7).

Again we introduce denotations 1)-5) from Section 2 induced now by the recurrent inequality (3.11):

1) $\psi^{-1}(z)$ and $\Phi^{-1}(z)$ are the inverse functions to $\psi(t)$ and $\Phi(t)$, respectively;

2) $v(s)=\psi^{-1}\left[C_1c_0\alpha(s-1)\right]+2C_1\alpha^2(s-1), c_0>1$ is a fixed control parameter;

3) $u(s,C_2)=\Phi^{-1}\left[\Phi(C_2)-a\left(F(s)-F(2)\right)\right], C_2>0,\ a=\dfrac{c_0-1}{c_0}<1$

4) $w(s,v(r))=\Phi^{-1}\left[\Phi(v(r))-a\left(F(s)-F(2)\right)\right],$

$2\le s,r<\infty;$

5) $Q=\Phi^{-1}\left[\Phi\left(\left\|x_1-x^*\right\|^2\right)-a\left(F(2)-F(1)\right)\right].$

Introduce also the basic condition (P).

Theorem 4.1. Assume that all the conditions of Theorem 3.1 are fulfiled and

i) the condition (P) holds for the functions

$w(s,v(2))=u(s,v(2))$ and $v(s)$;

ii) $u(s, v(2)) \geq v(s)$ as $s \to \infty$;

iii) $c_0 > 1$ is chosen such that $u(s, v(2)) \geq v(s)$ as $s \to 2$.

Then the sequence $\{x_n\}$ generated by (1.5)-(1.7) converges in average to a unique fixed point x^* of T and for all $n \geq 1$

$$E\left[\left\|x_n - x^*\right\|^2\right] \leq C_0 u(n, C_2),$$
$$C_2 = \max\{Q, v(2)\}. \tag{4.1}$$

Theorem 4.2. Assume that all the conditions of Theorem 3.1 are fulfiled and i) the condition (P) holds for the functions $v(s)$ and $w(s, v(z))$ with any fixed $z \in [2, \infty)$;

ii) $u(s, v(2)) \leq v(s)$ as $s \to \infty$;

iii) If $Q \leq v(2)$ and $c_0 > 1$ is chosen such that $u(s, v(2)) \leq v(s)$ as $s \to 2$, then the sequence $\{x_n\}$ generated by (1.5)-(1.7) converges in average to a unique fixed point x^* of T and for all $n \geq 1$

$$E\left[\left\|x_n - x^*\right\|^2\right] \leq C_0 v(n); \tag{4.2}$$

iv) In all the remaining cases, (4.1) holds for $1 \le n \le \bar{s}$ and (4.2) for $n \ge \bar{s}$, where $\bar{s}$ is a unique root of the equation $u(s, C_2) = v(s)$ on the interval $[2, \infty)$.

Let us provide the examples of functions $\psi(\tau)$ and $\Phi(\lambda)$ suitable for Theorems 4.1 and 4.2 (see [12,13]).

1) Below in Corollaries 4.3-4.6 we use the functions $\psi(\tau) = \tau^{\nu}$ with $\nu \ge 1$. For them

$$\Phi(\lambda) = \begin{cases} \ln \lambda, \text{ if } \nu = 1, \\ \dfrac{\lambda^{1-\nu}}{1-\nu}, \text{ if } \nu \ne 1, \end{cases} \tag{4.3}$$

and

$$\Phi^{-1}(z) = \begin{cases} \exp z, \text{ if } \nu = 1, \\ \left[(1-\nu)z\right]^{\frac{1}{1-\nu}}, \text{ if } \nu \ne 1, \end{cases} \tag{4.4}$$

2) If $\psi(\tau) = \exp \tau - 1, \tau > 0,$ then

$\Phi(\tau) = \ln(1 - \exp \tau)$ and $\Phi^{-1}(z) = -\ln(1 - \exp z)$.

3) If $\psi(\tau) = \dfrac{\tau}{\tau+1}, \tau > 0,$ then

$\Phi(\tau) = \ln \tau + \tau$ and $\Phi^{-1}(z) \le \ln(1 + \exp z)$.

4) If $\psi(\tau) = \dfrac{\tau^2}{\tau+1}, \tau > 0,$ then

$\Phi(\tau) = \ln \tau - \dfrac{1}{\tau}$.

In this example we are unable to define $\Phi^{-1}(z)$ in analitical form, therefore suggest to calculate it numerically by computer.

We next present very important corollaries from Theorems 4.1 and 4.2, where their assumptions automatically guarantee accomplishment of the condition (P) (see [4]). The functions $\psi(\tau)$ coincide with the point 1) above.

Corollary 4.3. Assume that $T : G \to H$ is a strongly contractive mapping, that is, (1.4) is satisfied with

$0 < q < 1$. Let in (1.5) $\alpha_n = \frac{b}{n}, b > 0.$ Then

$$F(t) - b\ln t, \Phi(\tau) = \frac{\ln \tau}{q_1}, \Phi^{-1}(z) = e^{q_1 z}, q_1 = 1 - q,$$

$$v(s) = \left(\frac{2c_0C_1}{b} + 2C_1\right)\frac{1}{s-1}, u(s,C) = C\left(\frac{2}{s}\right)^{abq_1},$$

$$Q = \|x_1 - x^*\|^2 \left(\frac{1}{2}\right)^{abq_1}.$$

I. Suppose that $b > \frac{1}{q_1}$ and $c_0 \geq \frac{bq_1}{bq_1 - 1}$. Then

$\lim_{n\to\infty} E\left[\|x_n - x^*\|^2\right] = 0$ and 1) If $Q \leq v(2)$ and $c_0 \geq \frac{bq_1}{bq_1 - 2}$, we have for all $n > 1$

$$E\left[\|x_n - x^*\|^2\right] \leq v(n); \tag{4.5}$$

2) In all the remain cases

$$E\left[\|x_n - x^*\|^2\right] \leq u(n, C),$$
$$C = \max\{Q, v(2)\}, 1 < n \leq \bar{s}, \tag{4.6}$$

and

$$E\left[\left\|x_n - x^*\right\|^2\right] \le v(s),\ n \ge \bar{s}, \tag{4.7}$$

where $\bar{s}$ is a unique root of the equation

$u(s,C) = v(s)$ on the interval $[2,\infty)$.

II. Suppose that $\frac{1}{2q_1} < b < \frac{1}{q_1}$ and

Then and the estimate (4.6) holds for all

Corollary 4.4. Assume that is a strongly contractive mapping, that is, in (1.4) is satisfied with

$0 < q < 1$. Let in (1.5) $\alpha_n = \frac{b}{n^\kappa}, b > 0, 0 < \kappa < 1.$ Then

$$F(t) = \frac{b}{1-\kappa} t^{1-\kappa}, \Phi(\tau) = \frac{\ln \tau}{q_1},$$

$$\Phi^{-1}(z) = \left(\frac{1-\kappa}{b} z\right)^{\frac{1}{1-\kappa}}, q_1 = 1 - q,$$

$$v(s) = \left(\frac{2c_0 C_1}{b} + 2C_1\right)\frac{1}{s-1},$$

$$u(s,C) = C \exp\left\{-\frac{abq_1}{1-\kappa}\left(s^{1-\kappa} - 2^{1-\kappa}\right)\right\},$$

$$Q = \left\|x_1 - x^*\right\|^2 \exp\left\{-\frac{abq_1}{1-\kappa}\left(2^{1-\kappa} - 1\right)\right\}.$$

Suppose that $c_0 \ge \frac{bq}{bq-1}$. Then $\lim_{n\to\infty} E\left[\left\|x_n - x^*\right\|^2\right] = 0$

and 1) If $Q \le v(2)$ and $c_0 \ge \frac{bq_1}{bq_1 - 2}$, we have for all

$n > 1$

$$E\left[\left\|x_n - x^*\right\|^2\right] \le v(n); \tag{4.8}$$

2) In all the remain cases the estimates (4.6) and (4.7) hold.

Corollary 4.5 Assume that $T : G \to H$ is a weakly contractive mapping of the class $C_{\phi(t)} = t^{2\nu-1}, \nu > 1,$ that is, in Theorem 3.1 $\psi(\tau) = \tau^{\nu}$. Let in (1.5)

$\alpha_n = \frac{b}{n}, b > 0.$ Then

$$F(t) = b\ln t, \Phi(\tau) = \frac{\tau^{1-\nu}}{1-\nu}, \Phi^{-1}(z) = \left[(1-\nu)z\right]^{\frac{1}{1-\nu}},$$

$$u(s,C) = C\left[1 + (\nu-1)C^{\nu-1}ab\ln\frac{s}{2}\right]^{-\frac{1}{\nu-1}},$$

$$Q = \left\|x_1 - x^*\right\|^2\left[1 + (\nu-1)\left\|x_1 - x^*\right\|^{2(\nu-1)} ab\ln 2\right]^{-\frac{1}{\nu-1}}.$$

If $c_0 > 1$ is chosen from the condition

$$\left(\frac{2c_0C_1}{b} + 2C_1\right)^{\nu-1} ab \le \nu^{-1},$$

then $\lim_{n\to\infty} E\left[\left\|x_n - x^*\right\|^2\right] = 0$ and for all $n > 1$

$$E\left[\left\|x_n - x^*\right\|^2\right] \le u(n,C),$$
$$C = \max\left\{Q, \left(\frac{2C_1c_0}{b}\right)^{\frac{1}{\nu}} + 2C_1\right\}.$$

(4.9)

Corollary 4.6. Assume that $T:G\to H$ is a weakly contractive mapping of the class $C_{\phi(t)} = t^{2\nu-1}, \nu > 1,$ that is, in Theorem 3.1 $\psi(\tau) = \tau^{\nu}.$
Let in (1.5)

$\alpha_n = \frac{b}{n^{\kappa}}, b > 0, 0 < \kappa < 1.$ Then

$$F(t) = \frac{b}{1-\kappa}t^{1-\kappa}, \Phi(\tau) = \frac{\tau^{1-\nu}}{1-\nu}, \Phi^{-1}(z) = \left[(1-\nu)z\right]^{\frac{1}{1-\nu}},$$
$$v(s) = \left[\left(\frac{2c_0C_1}{b}\right)^{\frac{1}{\nu}} + 2C_1\right]\left(\frac{1}{s-1}\right)^{\frac{\kappa}{\nu}},$$
$$u(s,C) = C\left[1 + \frac{\nu-1}{1-\kappa}C^{\nu-1}ab\left(s^{1-\kappa} - 2^{1-\kappa}\right)\right]^{-\frac{1}{\nu-1}},$$
$$Q = \left\|x_1 - x^*\right\|^2\left[1 + \frac{\nu-1}{1-\kappa}\left\|x_1 - x^*\right\|^{2(\nu-1)}ab\left(2^{1-\kappa} - 1\right)\right]^{-\frac{1}{\nu-1}}.$$

I. Suppose that

$$\frac{\kappa}{\nu} < \frac{1-\kappa}{\nu-1}.$$

1) If $Q \le v(2)$ and $c_0 > 1$ is chosen from the condition

$$\left[\left(\frac{2c_0C_1}{b}\right)^{\frac{1}{\nu}}+2C_1\right]^{\nu-1} ab \geq \frac{2^{\kappa}\kappa}{\nu},$$

then $\lim_{n\to\infty} E\left[\left\|x_n - x^*\right\|^2\right]=0$ and for all $n>1$

$$E\left[\left\|x_n - x^*\right\|^2\right] \leq v(n). \tag{4.10}$$

2) In all the remain cases the estimates (4.6) and (4.7) hold.

II. Suppose that

$$\frac{\kappa}{\nu} \geq \frac{1-\kappa}{\nu-1}.$$

If $c_0 > 1$ is chosen from the condition

$$\left[\left(\frac{2c_0C_1}{b}\right)^{\frac{1}{\nu}}+2C_1\right]^{\nu-1} ab \leq \frac{\kappa}{\nu},$$

then $\lim_{n\to\infty} E\left[\left\|x_n - x^*\right\|^2\right]=0$ and for all $n \geq 1$

$$E\left[\left\|x_n - x^*\right\|^2\right] \leq u(n,C), C = \max\{Q, v(2)\}. \tag{4.11}$$

In addition to the examples presented in this section, we produce the functions and $\Phi(\lambda)$ which have $\tau = 0$ as a tangency point of the infinite degree multiplicity and given logarithmic estimates of the convergence rate.

We define the function $\psi(\tau)$ by the following way:

$$\psi(\tau) = -\exp(-f(\tau))[f'(\tau)]^{-1}, \tau > 0,$$

where $f(\tau)$ is differentiable and decreasing function,

$\lim_{\tau \to 0} f(\tau) = \infty,$ and

$$\psi^{(l)}(0) = \lim_{\tau \to 0+} \psi^{(l)}(\tau), l = 0, 1, 2, \cdots,$$

where (l) denote the derivative degrees of the function

$\psi(\tau), \psi^{(0)}(\tau) = \psi(\tau).$ It is easy to see that

$$\Phi(\tau) = -\exp f(\tau)$$

and

$$\Phi^{-1}(z) = f^{-1}(\ln(-z)).$$

In particulari) $f(\tau) = \frac{1}{\tau}, \psi(\tau) = \tau^2 \exp\left(-\frac{1}{\tau}\right).$ We have

$\Phi(\tau) = -\exp\frac{1}{\tau}$ and $\Phi^{-1}(z) = \frac{1}{\ln(-z)}.$ We have to verify that $\psi(\tau)$ is convex. In fact, it is true because

$$\frac{d^2\psi(\tau)}{d\tau} = \left(2 + \frac{2\tau + 1}{\tau^2}\right)\exp\left(-\frac{1}{\tau}\right) > 0, \forall \tau > 0.$$

Beside this, it is easy to see that $\psi(\tau) \le \tau,$ at least, on the interval $[0,1]$. In the next examples we leave to readers to check these properties.

ii) $f(\tau) = \frac{1}{\tau^s},\ s \ge 1,\ \psi(\tau) = \exp\left(-\frac{1}{\tau^s}\right)\tau^{s+1}.$

We have $\Phi(\tau) = -\exp\frac{1}{\tau^s}$ and $\Phi^{-1}(z) = (\ln(-z))^{-\frac{1}{s}}$.

iii) $f(\tau) = \exp\frac{1}{\tau^s}, \psi(\tau) = \exp\left(-\left[\frac{1}{\tau^s} + \exp\left(\frac{1}{\tau^s}\right)\right]\right)\tau^{s+1}$

We have

$\Phi(\tau) = -\exp\left(\exp\frac{1}{\tau^s}\right)$ and $\Phi^{-1}(z) = (\ln\ln(-z))^{-\frac{1}{s}}$.

5. ALMOST SURE CONVERGENCE OF STOCHASTIC APPROXIMATIONS FOR NONEXPANSIVE MAPPINGS

Consider next the almost surely convergence of stochastic approximations. First of all, we need the stochastic analogy of Lemma 2.5:

Lemma 5.1. Let $\{\alpha_k\}$ be sequences of non-negative real numbers and $\{\beta_k\}$ be sequence of random A_k measurable variables, a.s. nonnegative for all $k \geq 1$. Assume that

$$\sum_{k=1}^{\infty} \alpha_k = \infty \text{ and } \sum_{k=1}^{\infty} \alpha_k \beta_k < \infty.$$

If $\lim_{k\to\infty} \alpha_k = 0$ and there exists $c > 0$ such that for all $k \geq 1$

$$\left|\beta_{k+1} - E[\beta_k / A_n]\right| \leq c\alpha_k \text{ a.s.,} \tag{5.1}$$

then $\lim_{k\to\infty}\beta_k = 0$ a.s.

The proof can be provided by the scheme of nonstochastic case (see Proposition 2 in [8]) or as it was done in [5].

We need also the following lemma from [14] as applied to our case of Hilbert spaces (the concepts of modulus of convexity $\delta_B(\varepsilon)$ of Banach spaces B or Hilbert spaces H can be found in [15] and [16]).

Lemma 5.2. If $F = I - T$ with a nonexpansive mapping $T : G \to H$, then for all $x, y \in D(T)$,

$$(Fx - Fy, x - y) \geq R_1^2 \delta_H\left(\frac{\|Fx - Fy\|}{2R_1}\right),$$

where

$$R_1 = \sqrt{2^{-1}\left(\|x-y\|^2 + \|Tx - Ty\|^2\right)} \leq \|x - y\|.$$

If $\|x\| \leq R$ and $\|y\| \leq R$ with $x, y \in D(T)$, then $R_1 \leq 2R$ and

$$(Fx - Fy, x - y) \geq L^{-1}R^2 \delta_H\left(\frac{\|Fx - Fy\|}{4R}\right).$$

Theorem 5.3. Assume that a mapping $T : G \to H$ is nonexpansive and its fixed point set N is nonempty. If

(1.8) holds and $E\left[\|\xi_n\| / A_n\right] \leq C_0$, then the sequence

$\{x_n\}$ generated by (1.5)-(1.7) weakly almost surely converges to some $\tilde{x} \in N$.

Proof. Let $x^* \in N$. We next use Lemma 5.2 and the estimate (see [17], p. 49)

$$\delta_H(\varepsilon) \geq \frac{\varepsilon^2}{8}$$

to get

$$(x - Tx, x - x^*) \geq R_1^2 \delta_H\left(\frac{\|x - Tx\|}{2R_1}\right) \geq \frac{\|x - Tx\|^2}{32}.$$

In this case the inequality (3.3) implies

$$\begin{aligned} &\|x_{n+1} - x^*\|^2 \\ &\leq (1 + 8\alpha_n^2)\|x_n - x^*\|^2 - (32)^{-1}\|\alpha_n x_n - Tx_n\|^2 \\ &-2\alpha_n(\xi_n, x_n - x^*) + 2\alpha_n^2\|\xi_n\|^2. \end{aligned} \tag{5.2}$$

Similarly to (3.10), we have

$$\begin{aligned} &E\left[\|x_{n+1} - x^*\|^2 / A_n\right] \\ &\leq (1 + 8\alpha_n^2)\|x_n - x^*\|^2 - 2\alpha_n E\left[\|x_n - Tx_n\|^2 / A_n\right] \\ &- 2\alpha_n E\left[(\xi_n, x_n - x^*) / A_n\right] + 2\alpha_n^2 E\left[\|\xi_n\|^2 / A_n\right]. \end{aligned} \tag{5.3}$$

Denote $\lambda_n = E\left[\|x_n - x^*\|^2\right]$ and $\beta_n = E\left[\|x_n - Tx_n\|^2\right]$

and apply the unconditional expectation to both sides of (5.3). Then

$$\lambda_{n+1} \leq (1 + 8\alpha_n^2)\lambda_n - 2\alpha_n\beta_n + 2C_1\alpha_n^2. \tag{5.4}$$

It follows from this that

$$\lambda_{n+1} \leq (1 + 8\alpha_n^2)\lambda_n + 2C_1\alpha_n^2.$$

Since $\sum_1^\infty \alpha_n^2 < \infty$ and due to Lemma 2.1, we conclude that $\{\lambda_n\}$ is bounded. Consequently, $\{x_n\}$ is bounded a.s. that follows from the theory of convergent quasimartingales (see [5,18]).

We now need Lemma 5.1. It is not difficult to see that

$$\sum_1^\infty \alpha_n \beta_n = \sum_1^\infty \alpha_n E\left[\|x_n - Tx_n\|^2\right] < \infty. \tag{5.5}$$

The last gives us

$$\sum_1^\infty \alpha_n \|x_n - Tx_n\|^2 < \infty \text{ a.s.}$$

Next we evaluate the following difference:

$$\begin{aligned} &\|x_{n+1} - Tx_{n+1}\|^2 - \|x_n - Tx_n\|^2 \\ &\le \left(\|x_{n+1} - Tx_{n+1}\| + \|x_n - Tx_n\|\right) \\ &\times \left(\|x_{n+1} - Tx_{n+1}\| - \|x_n - Tx_n\|\right) \end{aligned}$$

It is easy to see that $\|x_n - Tx_n\|$ is bounded a.s. Indeed, since $\|x_n\| \le C_1$ a.s., there exists a constant $C_2 > 0$ such that

$$\|x_n - Tx_n\| = 2\|x_n - x^*\| \le C_2 \text{ a.s.}$$

Therefore

$$\|x_{n+1} - Tx_{n+1}\| + \|x_n - Tx_n\| \le 2C_2 \text{ a.s.}$$

It is obviously that

$$\begin{aligned}
&\|x_{n+1} - Tx_{n+1}\| - \|x_n - Tx_n\| \\
&\le \|(x_{n+1} - x_n) + (Tx_n - Tx_{n+1})\| \le 2\|x_{n+1} - x_n\| \\
&= 2\|\Pr_G x_n - \alpha_n (x_n - Tx_n + \xi_n) - \Pr_G x_n\| \\
&\le 2\alpha_n \|x_n - Tx_n\| + 2\alpha_n \|\xi_n\|.
\end{aligned}$$

Thus,

$$\begin{aligned}
&E\left[\|x_{n+1} - Tx_{n+1}\|^2 / A_n\right] - \|x_n - Tx_n\|^2 \\
&\le 4\alpha_n C_2^2 + 4\alpha_n C_2 E\left[\|\xi_n\| / A_n\right] \\
&\le 4C_2 (C_2 + C_0)\alpha_n \text{ a.s..}
\end{aligned}$$

By Lemma 5.1, $\|x_n - Tx_n\| \to 0$ a.s. as $n \to \infty.$

Since $\{x_n\}$ is bounded a.s., there is a subsequence $\{x_{n_k}\}$ weakly convergent to some point $\tilde{x}.$ Since G is convex and closed, consequently, weakly closed, we assert that $\tilde{x} \in G.$ It is known that a nonexpansive mapping T is weakly demiclosed, therefore, $\tilde{x} \in N$ a.s. Weak almost surely convergence of whole sequence $\{x_n\}$ is shown by the standard way [8]. □

Corollary 5.4. Assume that $T : G \to H$ is a weakly contractive mapping of the class $C_{\phi(t)}.$ If $\sum_1^\infty \alpha_n = \infty,$ $\sum_1^\infty \alpha_n^2 < \infty$ and $E\left[\|\xi_n\| / A_n\right] \le C_0,$ then the sequence

$\{x_n\}$ generated by (1.5)-(1.7) strongly almost surely converges to unique fixed point x^* of $T.$

Proof. We have from (3.4)

$$\|x_n - Tx_n\|\|x_n - x^*\| \geq \psi\left(\|x_n - x^*\|^2\right). \tag{5.6}$$

Since $\{x_n\}$ is bounded a.s. and $\|x_n - Tx_n\| \to 0$ a.s.

as $n \to \infty,$ we conclude that $\psi\left(\|x_n - x^*\|^2\right) \to 0$ a.s.

The proof follows due to the properties of the function $\psi.$ □

Remark 5.5. It is clear that all the results remain still valid for self-mappings $T : G \to G.$ However, in this case, unlike any deterministic situation, the algorithm (1.5)-(1.7) must use the projection operator Pr_G because the vector $v_n = x_n - \alpha_n S_n x_n$ not always belongs to G. If $T_n x_n = Tx_n + \xi_n \in G$ for all $n \geq 1$ and $0 < \alpha_n < 1,$ then (1.5) can be replaced by

$$x_{n+1} = (1 - \alpha_n)x_n + \alpha_n T_n x_n, n = 1, 2, \cdots, x_1 \in G.$$

REFERENCES

1. M. T. Wasan, "Stochastic Approximation," Cambridge University Press, Cambridge, 1969.

2. M. B. Nevelson and R. Z. Hasminsky, "Stochastic Approximation and Recursive Estimation," AMS Providence, Rhode Island, 1973.

3. Ya. Alber and S. Shilman, "Nonasymptotic Estimates of the Convergence Rate of Stochastic Iterative Algorithms," Automation and Remote Control, Vol. 42, 1981, pp. 32-41.

4. Ya. Alber and S. Shilman, “General Nonasymptotic Estimates of the Convergence Rate of Iterative Stochastic Algorithms,” USSR Computational Mathematics and Mathematical Physics, Vol. 25, No. 2, 1985, pp. 13-20.

5. K. Barty, J.-S. Roy and C. Strugarek, “Hilbert-Valued Perturbed Subgradient Algorithms,” Mathematics of Operation Research, Vol. 32, No. 3, 2007, pp. 551-562.

6. X. Chen and H. White, “Asymptotic Properties of Some Projection-Based Robbins-Monro Procedures in a Hilbert Space,” Studies in Nonlinear Dynamics & Econometrics, Vol. 6, No. 1, 2002, pp. 1-53.

7. Ya. I. Alber and A. N. Iusem, “Extension of Subgradient Techniques for Nonsmooth Optimization in Banach Spaces,” Set-Valued Analysis, Vol. 9, No. 4, 2001, pp. 315-335.

8. Ya. Alber, A. Iusem and M. Solodov, “On the Projection Subgradient Method for Nonsmooth Convex Optimization in a Hilbert Space,” Mathematical Programming, Vol. 81, No. 1, 1998, pp. 23-35.

9. A. P. Dempster, N. M. Laird and D. B. Rubin, “Maximal Likelihood from Incomplete Data via the EM Algorithm,” Journal of the Royal Statistical Society, Vol. 39, No. 1, 1977, pp. 185-197.

10. W. Rudin, “Real and Complex Analysis,” McGraw-Hill, New York, 1978.

11. Ya. I. Alber and S. Guerre-Delabriere, “Principle of Weakly Contractive Maps in Hilbert Spaces,” Operator Theory, Advances and Applications, Vol. 98, 1997, pp. 7- 22.

12. Ya. I. Alber, “Recurrence Relations and Variational Inequalities,” Soviet Mathematics Doklady, Vol. 27, 1983, pp. 511-517.

13. Ya. Alber, S. Guerre-Delabriere and L. Zelenko, “The Principle of Weakly Contractive Maps in Metric Spaces,” Communications on Applied Nonlinear Analysis, Vol. 5, No. 1, 1998, pp. 45-68.

14. Ya. I. Alber, "New Results in Fixed Point Theory," Preprint, Haifa Technion, 2000.

15. J. Diestel, "The Geometry of Banach Spaces," Lecture Notes in Mathematics, No. 485, Springer, Berlin, 1975.

16. T. Figiel, "On the Moduli of Convexity and Smoothness," Studia Mathematica, Vol. 56, No. 2, 1976, pp. 121-155.

17. Ya. Alber and I. Ryazantseva, "Nonlinear Ill-Posed Problems of Monotone Type," Springer, Berlin, 2006.

18. M. Métivier, "Semimartingales," De Gruyter, Berlin, 1982.

7

CHAPTER

Deterministic Chaos of N Stochastic Waves in Two Dimensions

Victor A. Miroshnikov

Department of Mathematics, College of Mount Saint Vincent, New York, USA

ABSTRACT

Kinematic exponential Fourier (KEF) structures, dynamic exponential (DEF) Fourier structures, and KEF-DEF structures with time-dependent structural coefficients are developed to examine kinematic and dynamic problems for a deterministic chaos of N stochastic waves in the two-dimen- sional theory of the Newtonian flows with harmonic velocity. The Dirichlet problems are formulated for kinematic and dynamics systems of the vorticity, continuity, Helmholtz, Lamb-Helmholtz, and Bernoulli equations in the upper and lower domains for stochastic waves vanishing at infinity. Development of the novel method of solving partial differential equations through decomposition in invariant structures is resumed by using experimental and theoretical computation in Maple™. This computational method generalizes the analytical methods of separation of variables and undetermined coefficients. Exact solutions for the deterministic chaos of upper and lower cumulative flows are revealed by experimental computing, proved by theoretical computing, and justified by the system of Navier-Stokes PDEs. Various scenarios of a developed wave chaos are modeled by 3N parameters and 2N boundary functions, which exhibit stochastic behavior.

Keywords: Stochastic Waves, Invariant Structures, Experimental Computation, Theoretical Computation

1. INTRODUCTION

The two-dimensional (2d) Navier-Stokes system of partial differential equations (PDEs) for a Newtonian fluid with a constant density ρ and a constant kinematic viscosity ν in a gravity field g is

$$\frac{\partial \boldsymbol{v}}{\partial t}+(\boldsymbol{v}\cdot\nabla)\boldsymbol{v}=-\frac{1}{\rho}\nabla p_t+\nu\Delta\boldsymbol{v}+\boldsymbol{g}, \quad (1)$$

$$\nabla\cdot\boldsymbol{v}=0, \quad (2)$$

where $\boldsymbol{v}=(u,0,w)$ is a vector field of the flow velocity, $\boldsymbol{g}=(0,0,-g_z)$ is a vector field of the gravitational acceleration, p_t is a scalar field of the total pressure, $\nabla=(\partial/\partial x,0,\partial/\partial z)$ and $\Delta=\partial^2/\partial x^2+\partial^2/\partial z^2$ are the gradient and the Laplacian in the 2d Cartesian coordinate system $\boldsymbol{x}=(x,0,z)$ of the three-dimensional (3d) space with unit vectors $(\boldsymbol{i},\boldsymbol{j},\boldsymbol{k})$, respectively, and t is time.

By a flow vorticity $\boldsymbol{\omega}=(0,\upsilon,0)$ of the velocity field

$$\nabla\times\boldsymbol{v}=\boldsymbol{\omega}, \quad (3)$$

Equation (1) may be written into the Lamb-Pozrikidis form [1] [2]

$$\frac{\partial \boldsymbol{v}}{\partial t}+\nabla\left(\frac{1}{2}\boldsymbol{v}\cdot\boldsymbol{v}+\frac{p_t}{\rho}-\boldsymbol{g}\cdot\boldsymbol{x}\right)+\boldsymbol{\omega}\times\boldsymbol{v}+\nu\nabla\times\boldsymbol{\omega}=0, \tag{4}$$

which sets a dynamic balance of inertial, potential, vortical, and viscous forces, respectively.

Using a dynamic pressure per unit mass [3]

$$p_d=\frac{p_t-p_0}{\rho}-\boldsymbol{g}\cdot\boldsymbol{x}, \tag{5}$$

where p_0 is a reference pressure, a kinetic energy per unit mass $k_e=\boldsymbol{v}\cdot\boldsymbol{v}/2$, the 2d Helmholtz decomposition [4] of the velocity field

$$\boldsymbol{v}=\nabla\phi+\nabla\times\boldsymbol{\psi}, \tag{6}$$

and the vortex force

$$\boldsymbol{\omega}\times\boldsymbol{v}=\nabla d+\nabla\times\boldsymbol{a}, \tag{7}$$

Equation (4) is reduced to the Lamb-Helmholtz PDE [5]

$$\nabla b_e+\nabla\times\boldsymbol{h}_e=0 \tag{8}$$

for a scalar Bernoulli potential and a vector Helmholtz potential, respectively,

$$b_e=\frac{\partial\phi}{\partial t}+p_d+k_e+d, \tag{9}$$

$$h_e = \frac{\partial \psi}{\partial t} + \nu \omega + a, \tag{10}$$

where ϕ and d are scalar potentials, $\psi = (0, \eta, 0)$ and $a = (0, b, 0)$ are vector potentials, η and b are pseudovector potentials of ν and $\omega \times \nu$, respectively. The Lamb-Helmholtz PDE (8) means a dynamic balance between potential and vortical forces of the Navier-Stokes PDE (1), which are separated completely. Reduction of (1) to (8) means the potential-vortical duality of the Navier-Stokes PDE since writing equation (8) as

$$n_s = -\nabla b_e = \nabla \times h_e \tag{11}$$

shows that a virtual force n_s of (1) may be represented both in the potential form $n_s = -\nabla b_e$ and the vortical form $n_s = \nabla \times h_e$.

The exponential Fourier eigenfunctions obtained by the classical method of separation of variables of the 2d Laplace equation in [1] and [4] were primarily used for a linear part of the kinematic problem for free-surface waves of the theory of the ideal fluid with $\nu = 0$ in [6] . This analytical method was recently developed into the computational method of solving PDEs by decomposition into invariant structures. Topological flows away from boundaries were computed by the Boussinesq-Rayleigh-Taylor structures in [3] . Spatiotemporal cascades of exposed and hidden perturbations of the Couette flow were modeled by the trigonometric Taylor structures and the trigonometric-hyperbolic structures, respectively, in [7] . Dual perturbations of the Poiseuille-Hagen flow were treated by the invariant trigonometric, hyperbolic, and elliptic structures in [8] . Exact solutions for the conservative interaction of N internal waves were recently obtained by experimental and theoretical computing with kinematic Fourier (KF) structures with space-dependent structural coefficients and exponential kinematic Fourier (KEF) structures, dynamic exponential Fourier (DEF) structures, and KEF-DEF structures with constant structural coefficients in [5] .

To examine linear and nonlinear parts of kinematic and dynamic problems for 2d stochastic waves in the theory of Newtonian flows with harmonic velocity, the KEF structures, the DEF structures, and the KEF-DEF structures with time-dependent structural coefficients are developed in the current paper. The structure of this paper is as follows. The KEF structures are used to compute theoretical solutions of the kinematic problems for the velocity components and the dual potentials of the velocity field in Section 2. The KEF and KEF-DEF structures are employed for theoretical computation of the dynamic problems for the Helmholtz and Bernoulli potentials, the kinetic energy, and the total pressure in Section 3. Verification of the experimental and theoretical solutions is also provided in Section 3. Various scenarios of a developed wave chaos are treated in Section 4. A summary of main results is given in Section 5.

2. KINEMATIC PROBLEMS FOR INTERNAL WAVES

The following solutions and admissible boundary conditions for the kinematic problems of Section 2 in the KEF and DEF structures with time-dependent coefficients were primarily computed via experimental programming techniques, which use lists of equations and expressions of Maple™ in the virtual environment of a global variable Eqs with 25 procedures of 600 code lines in total.

2.1. Formulation of Theoretical Kinematic Problems for the Velocity Field

Theoretical kinematic problems for harmonic velocity components $u = u(x,z,t)$ and $w = w(x,z,t)$ of a cumulative flow $\boldsymbol{v} = u\boldsymbol{i} + w\boldsymbol{k}$ of a Newtonian fluid are given by vanishing the y-component of the vorticity Equation (3) and the continuity Equation (2), respectively,

$$\frac{\partial u}{\partial z} - \frac{\partial w}{\partial x} = 0, \quad \frac{\partial u}{\partial x} + \frac{\partial w}{\partial z} = 0. \tag{12}$$

To consider a deterministic chaos of N internal, stochastic waves, the cumulative flow is decomposed into a superposition of local flows

$$u=\sum_{n=1}^{N}u_n(x,z,t),\quad w=\sum_{n=1}^{N}w_n(x,z,t), \tag{13}$$

such that the local vorticity and continuity equations are

$$\frac{\partial u_n}{\partial z}-\frac{\partial w_n}{\partial x}=0,\quad \frac{\partial u_n}{\partial x}+\frac{\partial w_n}{\partial z}=0. \tag{14}$$

where $n=1,2,\cdots,N$.

An upper cumulative flow is specified by the Dirichlet condition in the KF structure on a lower boundary $z=0$ of an upper domain $x\in(-\infty,\infty)$ and $z\in[0,\infty)$ (see Figure 1).

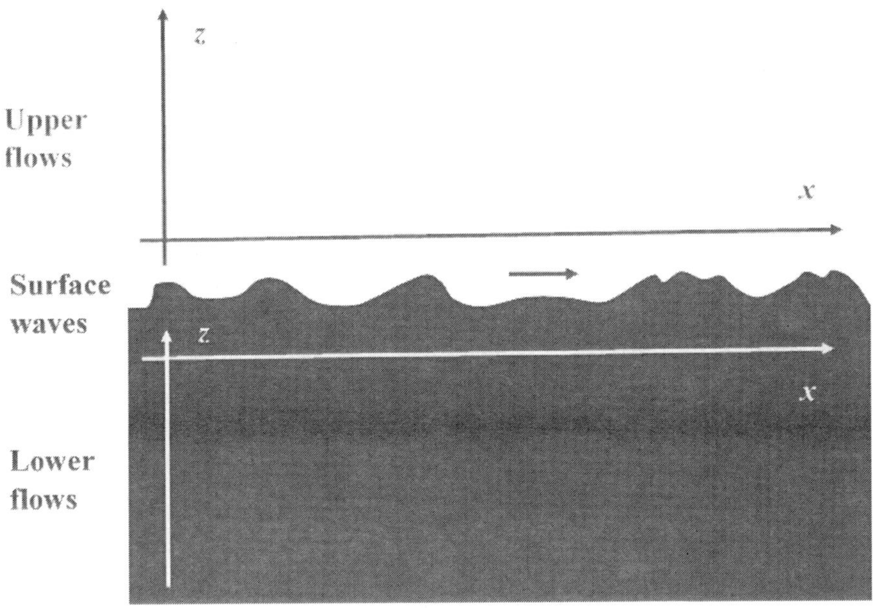

Figure 1. Configuration of upper and lower domains for stochastic waves.

$$w\big|_{z=0} = \sum_{n=1}^{N}\left[Fw_n(t)ca_n + Gw_n(t)sa_n\right], \tag{15}$$

and a vanishing condition as $z \to \infty$

$$w\big|_{z=\infty} = 0. \tag{16}$$

A lower cumulative flow is identified by the Dirichlet condition on an upper boundary $z=0$ of a lower domain $x \in (-\infty, \infty)$ and $z \in (-\infty, 0]$ (see Figure 1).

$$w\big|_{z=0} = \sum_{n=1}^{N}\left[Fw_n(t)ca_n + Gw_n(t)sa_n\right], \tag{17}$$

and a vanishing condition as $z \to \infty$

$$w\big|_{z=-\infty} = 0. \tag{18}$$

Thus, an effect of surface waves on the internal waves is described by the Dirichlet conditions (15) and (17). Here, a structural notation

$$ca_n = \cos(\alpha_n), \quad sa_n = \sin(\alpha_n), \tag{19}$$

is used for kinematic structural functions ca_n and sa_n, where $Fw_n(t)$ and $Gw_n(t)$ are time-dependent boundary functions, $\alpha_n = \rho_n X_n$ is an argument of the kinematic and dynamic structural functions, $X_n = x - Cx_n t - Xa_n$ is a propagation variable, ρ_n is a wave number, Cx_n is a celerity, and Xa_n is an initial coordinate for all n. Similar to [5] , boundary conditions for u_n are then redundant since boundary parameters of u_n depend on boundary parameters of w_n for the upper and lower flows, respectively, as

$$u|_{z=0} = -\sum_{n=1}^{N}\left[Gw_n(t)ca_n - Fw_n(t)sa_n\right], \tag{20}$$

$$u|_{z=0} = \sum_{n=1}^{N}\left[Gw_n(t)ca_n - Fw_n(t)sa_n\right]. \tag{21}$$

Similarly to w, u vanishes as $z \rightarrow \pm\infty$

$$u|_{z=\infty} = 0, \quad u|_{z=-\infty} = 0, \tag{22}$$

for the upper and lower cumulative flows, respectively.

2.2. Theoretical Solutions for the Velocity Field

Theoretical solutions of kinematic problems (12)-(18) are constructed in the KEF structure $p(x,z,t)$ of two spatial variables x, u and time t with a general term p_n, which in the structural notation may be written as

$$p(x,z,t) = \sum_{n=1}^{N} p_n(x,z,t) = \sum_{n=1}^{N}\left[fp_n(t)ca_n + gp_n(t)sa_n\right]\exp(\mp\rho_n z), \tag{23}$$

where signs "–" and "+" of the exponential term refer to the upper and lower flows, respectively, first letters f and g of structural coefficients $fp_n(t)$ and $gp_n(t)$ refer to the kinematic structural functions ca_n, sa_n and a second letter to the expanded variable p. General terms of the velocity components of the upper and lower flows become, respectively,

$$u_n = \left[fu_n(t) ca_n + gu_n(t) sa_n \right] \exp(\mp \rho_n z), \tag{24}$$

$$w_n = \left[fw_n(t) ca_n + gw_n(t) sa_n \right] \exp(\mp \rho_n z). \tag{25}$$

It may be shown that spatial derivatives of p_n are

$$\frac{\partial p_n}{\partial x} = \rho_n \left[gp_n(t) ca_n - fp_n(t) sa_n \right] \exp(\mp \rho_n z), \tag{26}$$

$$\frac{\partial p_n}{\partial z} = \mp \rho_n \left[fp_n(t) ca_n + gp_n(t) sa_n \right] \exp(\mp \rho_n z). \tag{27}$$

Application of (26) (27) to (24) (25), substitution in (14), collection of the structural functions, and vanishing their coefficients reduce two vorticity and continuity PDEs to the following system of four vorticity and continuity algebraic equations (AEs) with respect to $fu_n(t), gw_n(t), fw_n(t), gu_n(t)$ for the upper flows

$$-fu_n(t) - gw_n(t) = 0, \; fw_n(t) - gu_n(t) = 0, \tag{28}$$

$$-fu_n(t) - gw_n(t) = 0, \; -fw_n(t) + gu_n(t) = 0, \tag{29}$$

the lower flows

$$fu_n(t) - gw_n(t) = 0, \; fw_n(t) + gu_n(t) = 0, \tag{30}$$

$$-fu_n(t)+gw_n(t)=0,\ fw_n(t)+gu_n(t)=0, \tag{31}$$

and all $x, z, t, Xa_n, Cx_n, \rho_n, ca_n sa_n, fu_n(t), gu_n(t), fw_n(t), gw_n(t)$.

Solving AEs (28) and (30) yields for the upper and lower flows, respectively,

$$fu_n(t)=-gw_n(t),\ gu_n(t)=fw_n(t), \tag{32}$$

$$fu_n(t)=gw_n(t),\ gu_n(t)=-fw_n(t). \tag{33}$$

Substitution of solutions (32) in (29) and (33) in (31) reduces them to identities, showing that vorticity and continuity AEs (28)-(31) are compatible. Finally, substitution of (32) (33) in the KEF structures (24) (25) and solving the Dirichlet boundary conditions (15) and (17) with respect to $fw_n(t)$ and $gw_n(t)$ produces velocity components for the upper and lower cumulative flow, respectively,

$$u(x,z,t)=\mp\sum_{n=1}^{N}\left[Gw_n(t)ca_n - Fw_n(t)sa_n\right]\exp(\mp\rho_n z),$$
$$w(x,z,t)=\sum_{n=1}^{N}\left[Fw_n(t)ca_n + Gw_n(t)sa_n\right]\exp(\mp\rho_n z), \tag{34}$$

while vanishing boundary conditions (16) and (18) are obviously satisfied.

2.3. The DEF Structure and Theoretical Jacobian Determinants of the Velocity Field

Define two KEF structures $l(x,z,t)$ and $h(x,z,t)$ with general terms l_n and h_m by using the generalized Einstein notation for summation that is extended for exponents in [5]

$$l=\sum_{n=1}^{N} l_n=\left[fl_n(t)ca_n+gl_n(t)sa_n\right]\exp(\mp\rho_n z),$$
$$h=\sum_{m=1}^{N} h_m=\left[fh_m(t)ca_m+gh_m(t)sa_m\right]\exp(\mp\rho_n z). \tag{35}$$

Following [5] , define structural functions $Cad_{n,m}, Cas_{n,m}, Sad_{n,m}$, and $Sas_{n,m}$ of the DEF structure

$$Cad_{n,m}=\cos(\alpha_n-\alpha_m),\ \ Cas_{n,m}=\cos(\alpha_n+\alpha_m),$$
$$Sad_{n,m}=\sin(\alpha_n-\alpha_m),\ \ Sas_{n,m}=\sin(\alpha_n+\alpha_m), \tag{36}$$

where capital letters C and S stand for dynamic structural functions cosine and sine, letter a for arguments α_n, α_m, letters s and d for sum and difference of arguments α_n and α_m.

Computation of a general term $p_{n,n}=l_n h_n$ by summation of diagonal terms yields

$$p_{n,n}=\left\{fl_n(t)fh_n(t)+gl_n(t)gh_n(t)+\left[fl_n(t)fh_n(t)-gl_n(t)gh_n(t)\right]Cas_{n,n}\right.$$
$$\left.+\left[fl_n(t)gh_n(t)+fh_n(t)gl_n(t)\right]Sas_{n,n}\right\}\exp(\mp 2\rho_n z)/2. \tag{37}$$

A general term $p_{n,m}=l_n h_m$ computed by rectangular summation of non-diagonal terms becomes

$$p_{n,m}=\{[fl_n(t)fh_m(t)+gl_n(t)gh_m(t)]Cad_{n,m}+[fl_n(t)fh_m(t)-gl_n(t)gh_m(t)]Cas_{n,m}$$
$$+[-fl_n(t)gh_m(t)+fh_m(t)gl_n(t)]Sad_{n,m}+[fl_n(t)gh_m(t)+fh_m(t)gl_n(t)]Sas_{n,m}\}$$
$$\times\exp[\mp(\rho_n+\rho_m)z]/2. \tag{38}$$

By triangular summation, $p_{n,m}$ is reduced to

$$p_{n,m}=\{[fl_n(t)fh_m(t)+fl_m(t)fh_n(t)+gl_n(t)gh_m(t)+gl_m(t)gh_n(t)]Cad_{n,m}$$
$$+[fl_n(t)fh_m(t)+fl_m(t)fh_n(t)-gl_n(t)gh_m(t)-gl_m(t)gh_n(t)]Cas_{n,m}$$
$$+[-fl_n(t)gh_m(t)+fl_m(t)gh_n(t)-fh_n(t)gl_m(t)+fh_m(t)gl_n(t)]Sad_{n,m}$$
$$+[fl_n(t)gh_m(t)+fl_m(t)(t)gh_n(t)+fh_n(t)gl_m(t)+fh_m(t)gl_n(t)]Sas_{n,m}\}\exp[\mp(\rho_n+\rho_m)z]/2. \tag{39}$$

By (37) and (39), summation formula for the product of the KEF structures may be written as the DEF structure with time-dependent structural coefficients

$$p(x,z,t)=l(x,z,t)h(x,z,t)=\frac{1}{2}\sum_{n=1}^{N}[fdp_{n,n}(t)+fsp_{n,n}(t)Cas_{n,n}+gsp_{n,n}(t)Sas_{n,n}]\exp(\mp 2\rho_n z)$$
$$+\frac{1}{2}\sum_{n=1}^{N-1}\sum_{m=n+1}^{N}[fdp_{n,m}(t)Cad_{n,m}+fsp_{n,m}(t)Cas_{n,m}+gdp_{n,m}(t)Sad_{n,m}$$
$$+gsp_{n,m}(t)Sas_{n,m}\exp[\mp(\rho_n+\rho_m)z] \tag{40}$$

with the following structural coefficients:

$$fdp_{n,n}(t)=fl_n(t)fh_n(t)+gl_n(t)gh_n(t),$$
$$fsp_{n,n}(t)=fl_n(t)fh_n(t)-gl_n(t)gh_n(t),$$
$$gsp_{n,n}(t)=fl_n(t)gh_n(t)+fh_n(t)gl_n(t),$$
$$fdp_{n,m}(t)=fl_n(t)fh_m(t)+fl_m(t)fh_n(t)+gl_n(t)gh_m(t)+gl_m(t)gh_n(t),$$
$$fsp_{n,m}(t)=fl_n(t)fh_m(t)+fl_m(t)fh_n(t)-gl_n(t)gh_m(t)-gl_m(t)gh_n(t),$$
$$gdp_{n,m}(t)=-fl_n(t)gh_m(t)+fl_m(t)gh_n(t)-fh_n(t)gl_m(t)+fh_m(t)gl_n(t),$$
$$gsp_{n,m}(t)=fl_n(t)gh_m(t)+fl_m(t)gh_n(t)+fh_n(t)gl_m(t)+fh_m(t)gl_n(t), \tag{41}$$

where first two letters fd, fs, gd, and gs of structural coefficients $fdp_{n,m}, fsp_{n,m}, gdp_{n,m}$, and $gsp_{n,m}$ stand for dynamic structural functions Cad_n, Cas_n, Sad_n, and Sas_n, respectively, and a third letter for variable p.

Computation of local JDs for the velocity components of the upper and lower flow, respectively, yields

$$\frac{\partial u_n}{\partial x}\frac{\partial w_n}{\partial z}-\frac{\partial u_n}{\partial z}\frac{\partial w_n}{\partial x}=-\left[Fw_n^2(t)+Gw_n^2(t)\right]\rho_n^2\exp(\mp 2\rho_n z). \tag{42}$$

Thus, velocity components u_n and w_n are independent for non-trivial structural coefficients $Fw_n(t)$ and $Gw_n(t)$ since the local JDs vanish when $Fw_n^2(t)+Gw_n^2(t)=0$.

Computation of a global JD by using (40) (41) for velocity components of the upper and lower cumulative flows (34) with slant internal waves gives

$$\begin{aligned} J_g=\frac{\partial u}{\partial x}\frac{\partial w}{\partial z}-\frac{\partial u}{\partial z}\frac{\partial w}{\partial x}&=-\sum_{n=1}^{N}\left[Fw_n^2(t)+Gw_n^2(t)\right]\rho_n^2\exp(\mp 2\rho_n z)\\ &-2\sum_{n=1}^{N-1}\sum_{m=n+1}^{N}\{\left[Fw_n(t)Fw_m(t)+Gw_n(t)Gw_m(t)\right]Cad_{n,m}\\ &+\left[-Fw_n(t)Gw_m(t)+Fw_m(t)Gw_n(t)\right]Sad_{n,m}\}\rho_n\rho_m\exp\left[\mp(\rho_n+\rho_m)z\right]. \end{aligned} \tag{43}$$

So, J_g is a superposition of a propagation JD with general term $Jc_{n,n}$ proportional to $Cad_{n,n}\equiv 1$, an interaction JD with $Jc_{n,m}$ proportional to $Cad_{n,m}$, and an interaction JD with $Js_{n,m}$ proportional to $Sad_{n,m}$, which describe interaction between parallel and orthogonal internal waves, respectively.

$Jc_{n,n}$ coincides with local JDs (42). They describe propagation of internal waves and vanish only for vanishing waves with $Fw_n^2(t)+Gw_n^2(t)=0$. $Js_{n,m}$ vanishes for parallel waves with

$$\frac{Fw_m(t)}{Fw_n(t)}=\frac{Gw_m(t)}{Gw_n(t)}=A_{n,m}(t). \tag{44}$$

Global JD (43) then becomes

$$J_p=\sum_{n=1}^{N}\left[Fw_n^2(t)+Gw_n^2(t)\right]\rho_n^2\exp(\mp 2\rho_n z)$$
$$-2\sum_{n=1}^{N-1}\sum_{m=n+1}^{N}A_{n,m}(t)\left[Fw_n^2(t)+Gw_n^2(t)\right]Cad_{n,m}\rho_n\rho_m\exp\left[\mp(\rho_n+\rho_m)z\right]. \tag{45}$$

Thus, the global JD does not vanish for parallel waves with non-vanishing $Fw_n^2(t)+Gw_n^2(t)$.

$Jc_{n,m}$ vanishes for orthogonal waves with

$$\frac{Fw_m(t)}{Gw_n(t)}=-\frac{Gw_m(t)}{Fw_n(t)}=B_{n,m}(t). \tag{46}$$

In this case, global JD (43) is reduced to

$$J_o=-\sum_{n=1}^{N}\left[Fw_n^2(t)+Gw_n^2(t)\right]\rho_n^2\exp(\mp 2\rho_n z)$$
$$-2\sum_{n=1}^{N-1}\sum_{m=n+1}^{N}B_{n,m}(t)\left[Fw_n^2(t)+Gw_n^2(t)\right]Sad_{n,m}\rho_n\rho_m\exp\left[\mp(\rho_n+\rho_m)z\right]. \tag{47}$$

So, the global JD does not vanish also for orthogonal waves with non-vanishing $Fw_n^2(t)+Gw_n^2(t)$. In the general case of slant internal waves

(43), both $Js_{n,m}$ and $Jc_{n,m}$ are non-vanishing. Therefore, both propagating and interacting waves are independent for structural coefficients with $Fw_n^2(t)+Gw_n^2(t)\neq 0$ for all n.

2.4. Theoretical Solutions for the Kinematic Potentials in the KEF Structures

Theoretical kinematic problems for cumulative pseudovector potential $\eta(x,z,t)$ and cumulative scalar potential $\phi(x,z,t)$ of $\boldsymbol{v}$ are set by the global Helmholtz PDEs (6)

$$\frac{\partial\eta}{\partial z}+u=0,\quad \frac{\partial\eta}{\partial x}-w=0, \tag{48}$$

$$\frac{\partial\phi}{\partial x}-u=0,\quad \frac{\partial\phi}{\partial z}-w=0, \tag{49}$$

since the potential-vortical duality the velocity field admits two presentations: $v=\nabla\phi$ for $\psi=0$ and $v=\nabla\times\psi$ for $\phi=0$. The cumulative kinematic potentials are decomposed into a superposition of local kinematic potentials

$$\eta=\sum_{n=1}^{N}\eta_n(x,z,t),\quad \phi=\sum_{n=1}^{N}\phi_n(x,z,t), \tag{50}$$

such that the local Helmholtz PDEs are

$$\frac{\partial\eta_n}{\partial z}+u_n=0,\quad \frac{\partial\eta_n}{\partial x}-w_n=0, \tag{51}$$

$$\frac{\partial \phi_n}{\partial x} - u_n = 0, \quad \frac{\partial \phi_n}{\partial z} - w_n = 0, \tag{52}$$

where $n = 1, 2, \cdots, N$.

Construct general terms of the kinematic potentials of the local flows in the KEF structure with time-depen- dent coefficients

$$\eta_n = \left[fe_n(t) ca_n + ge_n(t) sa_n \right] \exp(\mp \rho_n z), \tag{53}$$

$$\varphi_n = \left[fp_n(t) ca_n + gp_n(t) sa_n \right] \exp(\mp \rho_n z). \tag{54}$$

Application of (26) (27) to (53) (54), substitution in the Helmholtz PDEs (51) (52), collection of the structural functions, and vanishing their coefficients reduce four Helmholtz PDEs to the following system of eight Helmholtz AEs for the upper flows

$$\begin{aligned} &-\rho_n fe_n(t) - Gw_n(t) = 0, \ \rho_n ge_n(t) - Fw_n(t) = 0, \\ &-\rho_n fp_n(t) - Fw_n(t) = 0, \ \rho_n gp_n(t) + Gw_n(t) = 0, \end{aligned} \tag{55}$$

$$\begin{aligned} &-\rho_n fe_n(t) - Gw_n(t) = 0, \ -\rho_n ge_n(t) + Fw_n(t) = 0, \\ &-\rho_n fp_n(t) - Fw_n(t) = 0, \ -\rho_n gp_n(t) - Gw_n(t) = 0, \end{aligned} \tag{56}$$

the lower flows

$$\begin{aligned} &-\rho_n fe_n(t) - Gw_n(t) = 0, \ \rho_n ge_n(t) - Fw_n(t)(t) = 0, \\ &-\rho_n fp_n(t) + Fw_n(t) = 0, \ \rho_n gp_n(t) - Gw_n(t) = 0, \end{aligned} \tag{57}$$

$$\rho_n fe_n(t)+Gw_n(t)=0,\ \rho_n ge_n(t)-Fw_n(t)=0,$$
$$\rho_n fp_n(t)-Fw_n(t)=0,\ \rho_n gp_n(t)-Gw_n(t)=0, \tag{58}$$

and all $x, z, t, Xa_n, Cx_n, \rho_n, ca_n, sa_n, fe_n(t), ge_n(t), fp_n(t), gp_n(t), Fw_n(t)$, and $Gw_n(t)$.

Solving AEs (55) and (57) with respect to time-dependent structural coefficients $fe_n(t), ge_n(t), fp_n(t)$, and $gp_n(t)$ gives for the upper and lower flows, respectively,

$$fe_n(t)=-\frac{Gw_n(t)}{\rho_n},\ ge_n(t)=\frac{Fw_n(t)}{\rho_n},$$
$$fp_n(t)=\mp\frac{Fw_n(t)}{\rho_n},\ gp_n(t)=\mp\frac{Gw_n(t)}{\rho_n}. \tag{59}$$

Substitution of solutions (59) in AEs (56) and (58) reduces them to identities. Finally, substitution of structural coefficients (59) in the KEF structures (53) (54) and superpositions (50) yields the cumulative kinematic potentials in the KEF structures for the upper and lower cumulative flows, respectively,

$$\eta(x,z,t)=-\sum_{n=1}^{N}\frac{Gw_n(t)ca_n-Fw_n(t)sa_n}{\rho_n}\exp(\mp\rho_n z),$$
$$\phi(x,z,t)=\mp\sum_{n=1}^{N}\frac{Fw_n(t)ca_n+Gw_n(t)sa_n}{\rho_n}\exp(\mp\rho_n z). \tag{60}$$

The theoretical solutions in the KEF and DEF structures for the kinematic problems of Section 2 were computed utilizing theoretical programming methods with symbolic general terms in the virtual environment of a global variable Eqt with 21 procedures of 522 Maple code lines in total. The theoretical formulas for velocity components (34), the products of the KEF structures (40) (41), and the kinematic potentials (60) of the upper and lower cumulative flows were justified by the correspondent experimental solutions for N = 1, 3, 10.

3. DYNAMIC PROBLEMS FOR INTERNAL WAVES

The following solutions for the dynamic problems of Section 3 in the KEF, DEF, and KEF-DEF structures were primarily computed by experimental programming with lists of equations and expressions in the virtual environment of the global variable Eqs with 18 procedures of 470 code lines in total.

3.1. Theoretical Solutions for the Dynamic Potentials in the KEF Structures

Theoretical dynamic problems in the KF structures for the Helmholtz and Bernoulli potentials of the cumulative flows are set by the Lamb-Helmholtz PDEs (8)-(10) in the vortical presentation with $\phi = 0$

$$\frac{\partial b_e}{\partial x} - \frac{\partial h_e}{\partial z} = 0, \quad \frac{\partial b_e}{\partial z} + \frac{\partial h_e}{\partial x} = 0, \quad h_e = \frac{\partial \eta}{\partial t}. \tag{61}$$

Equations (61) are complemented by the local Lamb-Helmholtz PDEs

$$\frac{\partial be_n}{\partial x} - \frac{\partial he_n}{\partial z} = 0, \quad \frac{\partial be_n}{\partial z} + \frac{\partial he_n}{\partial x} = 0, \quad he_n = \frac{\partial \eta_n}{\partial t}, \tag{62}$$

since the cumulative dynamic potentials are again decomposed into the local dynamic potentials

$$h_e = \sum_{n=1}^{N} he_n(x,z,t), \quad b_e = \sum_{n=1}^{N} be_n(x,z,t). \tag{63}$$

Construct a general term of the Bernoulli potential of the local flows in the KEF structure with time-depen- dent coefficients

$$be_n = \left[fb_n(t) ca_n + gb_n(t) sa_n \right] \exp(\mp \rho_n z). \tag{64}$$

Computation of the temporal derivative of η_n, application of (26) (27), substitution in (62), collection of the structural functions, and vanishing their coefficients reduce two Lamb-Helmholtz PDEs to the following system of four Lamb-Helmholtz AEs with respect to $fb_n(t)$ and $gb_n(t)$ for the upper flows

$$-\rho_n \left[fb_n(t) + Cx_n Gw_n(t) \right] + \frac{\mathrm{d}Fw_n}{\mathrm{d}t} = 0, \quad \rho_n \left[gb_n(t) - Cx_n Fw_n(t) \right] - \frac{\mathrm{d}Gw_n}{\mathrm{d}t} = 0, \tag{65}$$

$$-\rho_n \left[fb_n(t) + Cx_n Gw_n(t) \right] + \frac{\mathrm{d}Fw_n}{\mathrm{d}t} = 0, \quad -\rho_n \left[gb_n(t) - Cx_n Fw_n(t) \right] + \frac{\mathrm{d}Gw_n}{\mathrm{d}t} = 0, \tag{66}$$

and the lower flows

$$-\rho_n \left[fb_n(t) - Cx_n Gw_n(t) \right] - \frac{\mathrm{d}Fw_n}{\mathrm{d}t} = 0, \quad \rho_n \left[gb_n(t) + Cx_n Fw_n(t) \right] + \frac{\mathrm{d}Gw_n}{\mathrm{d}t} = 0, \tag{67}$$

$$\rho_n \left[fb_n(t) - Cx_n Gw_n(t) \right] + \frac{\mathrm{d}Fw_n}{\mathrm{d}t} = 0, \quad \rho_n \left[gb_n(t) + Cx_n Fw_n(t) \right] + \frac{\mathrm{d}Gw_n}{\mathrm{d}t} = 0, \tag{68}$$

and all variables, parameters, and functions $x, z, t, Xa_n, Cx_n, \rho_n, ca_n sa_n, fb_n(t), gb_n(t), Fw_n(t)$, and $Gw_n(t)$.

Solving AEs (65) and (67) for structural coefficients $fb_n(t)$ and $gb_n(t)$ yields for the upper and lower flows, respectively,

$$fb_n(t) = \mp\left[Cx_n Gw_n(t) - \frac{1}{\rho_n}\frac{dFw_n}{dt}\right], \quad gb_n(t) = \pm\left[Cx_n Fw_n(t) + \frac{1}{\rho_n}\frac{dGw_n}{dt}\right]. \tag{69}$$

Substitution of solutions (69) in AEs (66) and (68) reduces them to identities. Eventually, substitution of structural coefficients (69) in the KEF structure (64) and superpositions (63) returns the cumulative dynamic potentials in the KEF structures for the upper and lower cumulative flows, respectively,

$$he(x,z,t) = -\sum_{n=1}^{N}\left\{\left[Cx_n Fw_n(t) + \frac{1}{\rho_n}\frac{dGw_n}{dt}\right]ca_n + \left[Cx_n Gw_n(t) - \frac{1}{\rho_n}\frac{dFw_n}{dt}\right]sa_n\right\}\exp(\mp\rho_n z), \tag{70}$$

$$be(x,z,t) = \pm\sum_{n=1}^{N}\left\{-\left[Cx_n Gw_n(t) - \frac{1}{\rho_n}\frac{dFw_n}{dt}\right]ca_n + \left[Cx_n Fw_n(t) + \frac{1}{\rho_n}\frac{dGw_n}{dt}\right]sa_n\right\}\exp(\mp\rho_n z). \tag{71}$$

3.2. Theoretical Solutions for the Total Pressure in the KEF-DEF Structures

Theoretical dynamic problems in the KEF-DEF structures for k_e, p_d, and p_t of the cumulative flows are formulated by definition

$$k_e(x,z,t) = \frac{1}{2}\left[u(x,z,t)^2 + w(x,z,t)^2\right], \tag{72}$$

the Bernoulli Equation (9) with $\phi = 0$

$$p_d(x,z,t) = b_e(x,z,t) - k_e(x,z,t), \tag{73}$$

and the hydrostatic Equation (5)

$$p_t(x,z,t) = p_0 - \rho g_z z + \rho p_d(x,z,t), \tag{74}$$

where p_0 is the reference pressure at $z = 0.$

Computation of k_e by (40) (41) and (34) returns

$$k_e(x,z,t) = \frac{1}{2}\sum_{n=1}^{N}\left[Fw_n^2(t)+Gw_n^2(t)\right]\exp(\mp 2\rho_n z)+\sum_{n=1}^{N-1}\sum_{m=n+1}^{N}\{\left[Fw_n(t)Fw_m(t)+Gw_n(t)Gw_m(t)\right]Cad_{n,m}$$
$$+\left[Fw_m(t)Gw_n(t)-Fw_n(t)Gw_m(t)\right]Sad_{n,m}\}\exp\left[\mp(\rho_n+\rho_m)z\right] \tag{75}$$

for the upper and lower cumulative flows, respectively. Substitution of (71), (75), and (73) in (74) yields

$$p_t(x,z,t) = p_0 - \rho g_z z + \rho\Bigg(\pm\sum_{n=1}^{N}\left\{-\left[Cx_n Gw_n(t) - \frac{1}{\rho_n}\frac{\mathrm{d}Fw_n}{\mathrm{d}t}\right]ca_n + \left[Cx_n Fw_n(t) + \frac{1}{\rho_n}\frac{\mathrm{d}Gw_n}{\mathrm{d}t}\right]sa_n\right\}\exp(\mp\rho_n z)$$
$$-\frac{1}{2}\sum_{n=1}^{N}\left[Fw_n^2(t)+Gw_n^2(t)\right]\exp(\mp 2\rho_n z)-\sum_{n=1}^{N-1}\sum_{m=n+1}^{N}\{\left[Fw_n(t)Fw_m(t)+Gw_n(t)Gw_m(t)\right]Cad_{n,m}$$
$$+\left[Fw_m(t)Gw_n(t)-Fw_n(t)Gw_m(t)\right]Sad_{n,m}\}\exp\left[\mp(\rho_n+\rho_m)z\right]\Bigg) \tag{76}$$

for the upper and lower cumulative flows, respectively. So, the kinetic energy is obtained in the DEF structures, the dynamic pressure is expressed in the KEF-DEF structures, and the total pressure is computed in the KEF-DEF and polynomial structures.

3.3. Harmonic Relationships between the Kinematic and Dynamic Variables

Similar to the invariant trigonometric, hyperbolic, and elliptic structures [8] , there are two pairs of independent KEF structures: generating structures with general terms As_n, Cs_n and complementary structures with general terms Bs_n, Ds_n for the upper and lower flows, respectively,

$$As_n = \left[Fw_n(t)ca_n + Gw_n(t)sa_n\right]\exp(\mp\rho_n z),\quad Bs_n = \left[Gw_n(t)ca_n - Fw_n(t)sa_n\right]\exp(\mp\rho_n z),\tag{77}$$

$$Cs_n = \left(\frac{\mathrm{d}Fw_n}{\mathrm{d}t}ca_n + \frac{\mathrm{d}Gw_n}{\mathrm{d}t}sa_n\right)\exp(\mp\rho_n z),\quad Ds_n = \left(\frac{\mathrm{d}Gw_n}{\mathrm{d}t}ca_n - \frac{\mathrm{d}Fw_n}{\mathrm{d}t}sa_n\right)\exp(\mp\rho_n z).\tag{78}$$

Expressing velocity components (34), kinematic potentials (60), and dynamic potentials (70) (71) through the generating and complementary structures (77) (78) and solving for As_n and Bs_n gives algebraic relationships

$$As_n = w_n = \mp\rho_n\varphi_n = -\frac{he_n}{Cx_n} - \frac{Ds_n}{\rho_n Cx_n},\quad Bs_n = \mp u_n = -\rho_n\eta_n = \mp\frac{be_n}{Cx_n} + \frac{Cs_n}{\rho_n Cx_n}.\tag{79}$$

Taking derivatives of (34) and (60) with respect to x, z and solving for As_n and Bs_n yields differential relationships, which extend algebraic ones (79),

$$As_n = \pm\frac{1}{\rho_n}\frac{\partial u_n}{\partial x} = \mp\frac{1}{\rho_n}\frac{\partial w_n}{\partial z} = \frac{\partial \eta_n}{\partial x} = \frac{\partial \varphi_n}{\partial z},\quad Bs_n = \frac{1}{\rho_n}\frac{\partial w_n}{\partial x} = \frac{1}{\rho_n}\frac{\partial u_n}{\partial z} = \mp\frac{\partial \varphi_n}{\partial x} = \pm\frac{\partial \eta_n}{\partial z}.\tag{80}$$

In fluid dynamics, relationships (79) (80) mean that a harmonic flow, which is non-uniform in $x-$ or $z-$ directions, produces a complementary flow in z- or x-directions, respectively.

Computing velocity components (34) and dynamic potentials (70) (71) through the generating and complementary structures (77) (78), taking temporal and spatial derivatives, and solving for be_n and he_n returns

$$\rho_n he_n = \pm\frac{\partial u_n}{\partial t} = \mp\frac{\partial be_n}{\partial x} = \mp\frac{\partial he_n}{\partial z}, \quad \rho_n be_n = \pm\frac{\partial w_n}{\partial t} = \pm\frac{\partial he_n}{\partial x} = \mp\frac{\partial be_n}{\partial z}. \tag{81}$$

Differential relationships (81) mean that spatial derivatives of the dynamic potentials generate temporal rates of a harmonic flow.

By the following substitutions:

$$u_n = u = \eta_n = \eta = he_n = he = u_c, \quad -w_n = -w = \phi_n = \phi = -be_n = -be = v_c, \tag{82}$$

both local and global vorticity and continuity Equations (12) and (14), Helmholtz Equations (48)-(49) and (51)- (52), and Lamb-Helmholtz Equations (61) and (62), respectively, are reduced to the Cauchy-Riemann equations

$$\frac{\partial u_c}{\partial x} - \frac{\partial v_c}{\partial z} = 0, \quad \frac{\partial u_c}{\partial z} + \frac{\partial v_c}{\partial x} = 0 \tag{83}$$

for conjugate functions u_c and v_c [4] . All conjugate functions have orthogonal isocurves, due to the vanishing scalar product of gradients,

$$\nabla u_c \cdot \nabla v_c = \frac{\partial u_c}{\partial x}\frac{\partial v_c}{\partial x} + \frac{\partial u_c}{\partial z}\frac{\partial v_c}{\partial z} = 0, \tag{84}$$

and vanishing Laplacians,

$$\frac{\partial^2 u_c}{\partial x^2} + \frac{\partial^2 u_c}{\partial z^2} = 0, \quad \frac{\partial^2 v_c}{\partial x^2} + \frac{\partial^2 v_c}{\partial z^2} = 0. \tag{85}$$

Thus, $(u_n, w_n), (u, w), (\eta_n, \phi_n), (\eta, \phi), (he_n, be_n)$,

and (he, be) are six pairs of harmonic functions with orthogonal isocurves.

3.4. Theoretical Verification by the System of Navier-Stokes PDEs

The system of the Navier-Stokes PDEs (1)-(2) in the scalar notation becomes

$$\frac{\partial u}{\partial t}+u\frac{\partial u}{\partial x}+w\frac{\partial u}{\partial z}=-\frac{1}{\rho}\frac{\partial p_t}{\partial x}+\nu\left(\frac{\partial^2 u}{\partial x^2}+\frac{\partial^2 u}{\partial z^2}\right),\quad \frac{\partial w}{\partial t}+u\frac{\partial w}{\partial x}+w\frac{\partial w}{\partial z}=-\frac{1}{\rho}\frac{\partial p_t}{\partial z}+\nu\left(\frac{\partial^2 w}{\partial x^2}+\frac{\partial^2 w}{\partial z^2}\right)-g_z, \tag{86}$$

$$\frac{\partial u}{\partial x}+\frac{\partial w}{\partial z}=0. \tag{87}$$

Computation of spatial derivatives of (34) by (26)-(27) reduces (87) to identity both for the upper and lower cumulative flows.

Temporal derivatives of $\boldsymbol{v}$ in the KEF structures for the upper and lower cumulative flows, respectively, are

$$\frac{\partial u}{\partial t}=\mp\sum_{n=1}^{N}\left\{\left[\rho_n Cx_n Fw_n(t)+\frac{\mathrm{d}Gw_n}{\mathrm{d}t}\right]ca_n+\left[\rho_n Cx_n Gw_n(t)-\frac{\mathrm{d}Fw_n}{\mathrm{d}t}\right]sa_n\right\}\exp(\mp\rho_n z), \tag{88}$$

$$\frac{\partial w}{\partial t}=\sum_{n=1}^{N}\left\{-\left[\rho_n Cx_n Gw_n(t)-\frac{\mathrm{d}Fw_n}{\mathrm{d}t}\right]ca_n+\left[\rho_n Cx_n Fw_n(t)+\frac{\mathrm{d}Gw_n}{\mathrm{d}t}\right]sa_n\right\}\exp(\mp\rho_n z). \tag{89}$$

The directional derivatives of (86) computed by (40)-(41) in the DEF structures for the upper and lower cumulative flows, respectively, become

$$(\boldsymbol{v}\cdot\nabla)u=\sum_{n=1}^{N-1}\sum_{m=n+1}^{N}\{[Fw_n(t)Gw_m(t)-Fw_m(t)Gw_n(t)]Cad_{n,m}$$
$$+[Fw_n(t)Fw_m(t)+Gw_n(t)Gw_m(t)]Sad_{n,m}\}(\rho_m-\rho_n)\exp[\mp(\rho_n+\rho_m)z],\tag{90}$$

$$(\boldsymbol{v}\cdot\nabla)w=\mp\sum_{n=1}^{N}[Fw_n^2(t)+Gw_n^2(t)]\exp(\mp2\rho_nz)\mp\sum_{n=1}^{N-1}\sum_{m=n+1}^{N}\{[Fw_n(t)Fw_m(t)+Gw_n(t)Gw_m(t)]Cad_{n,m}$$
$$-[Fw_n(t)Gw_m(t)-Fw_m(t)Gw_n(t)]Sad_{n,m}\}(\rho_m+\rho_n)\exp[\mp(\rho_n+\rho_m)z].\tag{91}$$

By using (26)-(27), components of the gradient of (76) may be written in the KEF-DEF structures for the upper and lower cumulative flows, respectively, as

$$\frac{\partial p_t}{\partial x}=\rho\Bigg(\pm\sum_{n=1}^{N}\left\{\left[\rho_nCx_nFw_n(t)+\frac{\mathrm{d}Gw_n}{\mathrm{d}t}\right]ca_n+\left[\rho_nCx_nGw_n(t)-\frac{\mathrm{d}Fw_n}{\mathrm{d}t}\right]sa_n\right\}\exp(\mp\rho_nz)$$
$$-\sum_{n=1}^{N-1}\sum_{m=n+1}^{N}\{[Fw_n(t)Gw_m(t)-Fw_m(t)Gw_n(t)]Cad_{n,m}$$
$$+[Fw_n(t)Fw_m(t)+Gw_n(t)Gw_m(t)]Sad_{n,m}\}(\rho_m-\rho_n)\exp[\mp(\rho_n+\rho_m)z]\Bigg),\tag{92}$$

$$\frac{\partial p_t}{\partial z}=\rho\Bigg(-g_z+\sum_{n=1}^{N}\left\{\left[\rho_nCx_nGw_n(t)-\frac{\mathrm{d}Fw_n}{\mathrm{d}t}\right]ca_n-\left[\rho_nCx_nFw_n(t)+\frac{\mathrm{d}Gw_n}{\mathrm{d}t}\right]sa_n\right\}\exp(\mp\rho_nz)$$
$$\pm\sum_{n=1}^{N}[Fw_n^2(t)+Gw_n^2(t)]\rho_n\exp(\mp2\rho_nz)\pm\sum_{n=1}^{N-1}\sum_{m=n+1}^{N}\{[Fw_n(t)Fw_m(t)+Gw_n(t)Gw_m(t)]Cad_{n,m}$$
$$-[Fw_n(t)Gw_m(t)-Fw_m(t)Gw_n(t)]Sad_{n,m}\}(\rho_m+\rho_n)\exp[\mp(\rho_n+\rho_m)z]\Bigg).\tag{93}$$

Substitution of Equations (88)-(93) and (85) for u and w in (86) reduces then to identities. Thus, Equations (34) and (76) constitute exact solutions in the KEF, DEF, and KEF-DEF structures with time-dependent structural coefficients for deterministic chaos of N stochastic waves both in the upper and lower domains.

The theoretical solutions in the KEF, DEF, and KEF-DEF structures for the dynamic problems of Section 3 were computed by theoretical programming methods with symbolic general terms in the virtual environment of the global variable Eqt with 14 procedures of 410 code lines in total. The theoretical solutions for Helmholtz and Bernoulli potentials (70)-(71), total pressure (76), temporal derivatives (88)-(89), directional derivatives (90)- (91), and pressure gradient (92)-(93) of the upper and lower cumulative flows were justified by the correspondent experimental solutions for N = 1, 3, 10.

All kinematic solutions of Section 2 and dynamic solutions of Section 3 with time-dependent structural coefficients are reduced to the correspondent solutions of [5] when $Fw_n(t) = Fw_n$ and $Gw_n(t) = Gw_n$.

4. CHAOTIC SCENARIOS

Boundary functions $Fw_n(t)$ and $Gw_n(t)$, which are shown in Figure 2 for $n = 3$, are constructed from stochastic solutions of the Navier-Stokes equation in one dimension by the following method. Let $Fd_n(x_d, \tau)$ and $Gd_n(x_d, \tau)$ be the stochastic solutions of

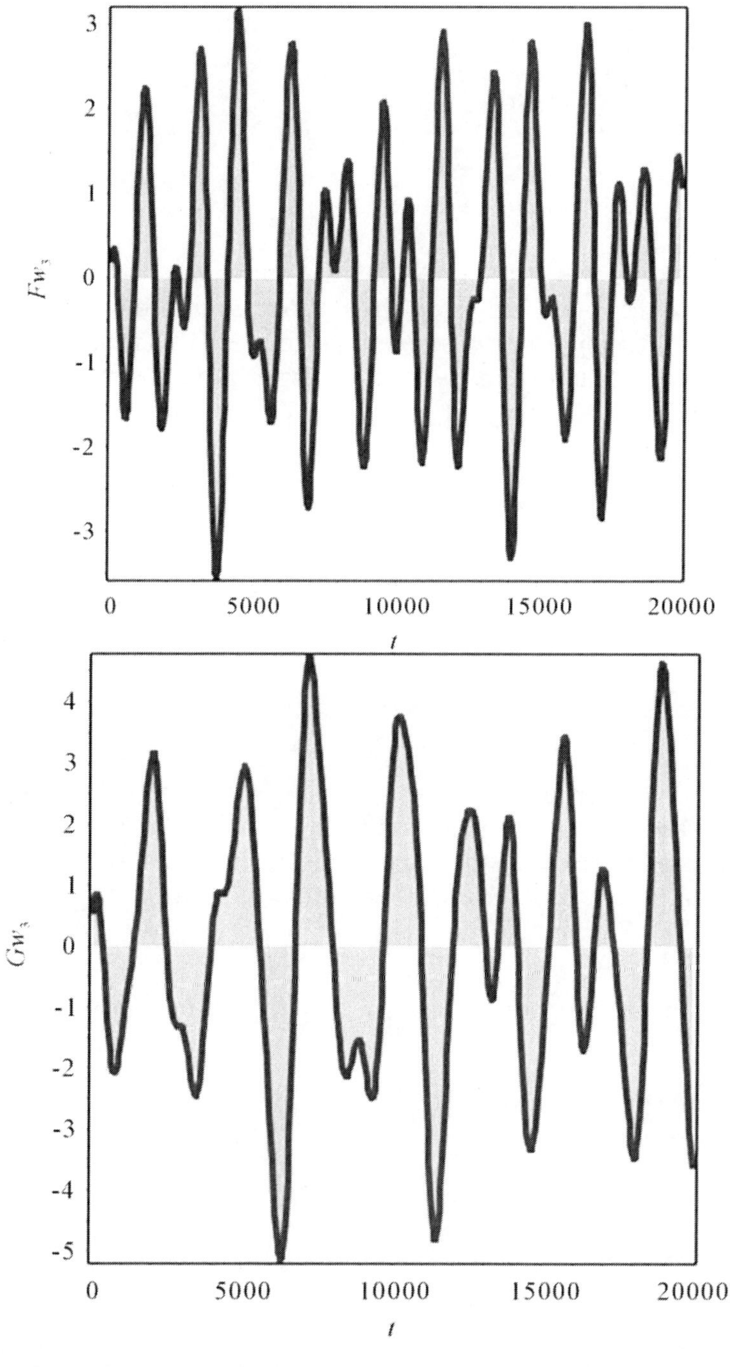

Figure 2. Stochastic boundary functions.

$$R_n \frac{\partial Fd_n}{\partial \tau} - \frac{\partial^2 Fd_n}{\partial x_d^2} = 0, \quad R_n \frac{\partial Gd_n}{\partial \tau} - \frac{\partial^2 Gd_n}{\partial x_d^2} = 0, \tag{94}$$

where R_n is the Reynolds number.

In agreement with [7] , $Fd_n(x_d, \tau)$ is composed of an invariant structure with even indices of hyperbolic secant-tangent temporal modes and trigonometric sine spatial modes (HETO structure) and an invariant structure of the same temporal modes combined with trigonometric cosine spatial modes (HETE structure)

$$\begin{aligned} Fd_n &= \sum_{m=1}^{n} \left(th_{n,m} \sum_{k=1}^{K} gd_{n,2k,m} sh_{n,2k,m} + \sum_{k=1}^{K} fd_{n,2k,m} sh_{n,2k,m} \right), \\ gd_{n,2k,m} &= \sum_{l=1}^{k} a_{2k,2l} \left(A_{n,2l,m} sn_{n,2l,m} + B_{n,2l,m} cs_{n,2l,m} \right), \\ fd_{n,2k,m} &= \sum_{l=1}^{k} b_{2k,2l} \left(A_{n,2l,m} sn_{n,2l,m} + B_{n,2l,m} cs_{n,2l,m} \right). \end{aligned} \tag{95}$$

$Gd_n(x_d, \tau)$ is formed of an invariant structure with odd indices of hyperbolic secant-tangent temporal modes and trigonometric sine spatial modes (HOTO structure) and an invariant structure of the same temporal modes composed with trigonometric cosine spatial modes (HOTE structure)

$$\begin{aligned} Gd_n &= \sum_{m=1}^{n} \left(th_{n,m} \sum_{k=1}^{K} gd_{n,2k-1,m} sh_{n,2k-1,m} + \sum_{k=1}^{K} fd_{n,2k-1,m} sh_{n,2k-1,m} \right), \\ gd_{n,2k-1,m} &= \sum_{k=1}^{K} a_{2k-1,2l-1} \left(A_{n,2l-1,m} sn_{n,2l-1,m} + B_{n,2l-1,m} cs_{n,2l-1,m} \right), \\ fd_{n,2k-1,m} &= \sum_{l=1}^{k} b_{2k-1,2l-1} \left(A_{n,2l-1,m} sn_{n,2l-1,m} + B_{n,2l-1,m} cs_{n,2l-1,m} \right). \end{aligned} \tag{96}$$

Here, a structural notation

$$th_{n,m} = \tanh\left(\delta_{n,m}\right),\ sh_{n,2k,m} = \operatorname{sech}^{2k}\left(\delta_{n,m}\right),\ sh_{n,2k-1,m} = \operatorname{sech}^{2k-1}\left(\delta_{n,m}\right),\ \delta_{n,m} = \omega_{n,m}\left(\tau - \tau_{n,m}\right),$$
$$sn_{n,2l,m} = \sin\left(\kappa_{n,2l,m}x_d\right),\ cs_{n,2l,m} = \cos\left(\kappa_{n,2l,m}x_d\right),\ sn_{n,2l-1,m} = \sin\left(\kappa_{n,2l,m}x_d\right),$$
$$cs_{n,2l-1,m} = \cos\left(\kappa_{n,2l,m}x_d\right),\ \kappa_{n,2l,m} = \sqrt{R_n\omega_{n,m}2l},\ \kappa_{n,2l-1,m} = \sqrt{R_n\omega_{n,m}\left(2l-1\right)} \tag{97}$$

is used for structural functions of temporal and spatial modes, where $m = 1,2,\cdots,n$ is an index of spatiotemporal components, $k = 1,2,\cdots,K$ is an index of temporal modes, $l = 1,2,\cdots,k$ is an index of spatial modes, $\omega_{n,m}$ is a frequency, $\tau_{n,m}$ is a delay parameter, $A_{n,2l-1,m}$ $B_{n,2l,m}$, $A_{n,2l,m}$, and $B_{n,2l-1,m}$ are spatial amplitudes. Structural coefficients $a_{2k,2l}$, $b_{2k,2l}$, $a_{2k-1,2l-1}$, and $b_{2k-1,2l-1}$ are computed by initialization conditions

$$a_{2k,2l} = 1,\ b_{2k,2l} = 1,\ a_{2k-1,2l-1} = 1,\ b_{2k-1,2l-1} = 1, \tag{98}$$

and recurrent relations

$$b_{2k,2l} = \frac{(2k-1)(k-1)}{2(k-l)(k+l)}b_{2k-2,2l},\ b_{2k-1,2l-1} = \frac{(2k-3)(k-1)}{2(k-l)(k+l-1)}b_{2k-3,2l-1},$$
$$a_{2k,2l} = \frac{k}{l}b_{2k,2l},\qquad a_{2k-1,2l-1} = \frac{2k-1}{2l-1}b_{2k-1,2l-1}. \tag{99}$$

In Figure 2, stochastic boundary functions

$$Fw_n\left(t\right) = Fd_n\left(t, Xa_nT_0\right),\quad Gw_n\left(t\right) = Gd_n\left(t, Xa_nT_0\right) \tag{100}$$

were evaluated for $N=3$, $K=5$, $R_n=10n^2$, temporal delays $\tau_{n,m}=T_0 m$, a scale of temporal delays $T_0=10^6$, $Xa_n=(3/2,1,1/2)$, scales of spatiotemporal components

$$C_h=1,\quad C_l=-12/5,\quad C_m=C_h+C_l\left[1+(-1)^m\right]/2 \tag{101}$$

and spatial amplitudes

$$A_{n,2l,m}=B_{n,2l,m}=C_m/(2l)!,\quad A_{n,2l-1,m}=B_{n,2l-1,m}=C_m/(2l-1)!, \tag{102}$$

which insure convergence of the spatial modes. Frequencies of $Fd_n(x_d,\tau)$:

$$\omega_{1,1}=1.847\times10^{-6},\ \omega_{2,1}=4.617\times10^{-7},\ \omega_{2,2}=1.649\times10^{-7},$$
$$\omega_{3,1}=2.052\times10^{-7},\ \omega_{3,2}=7.329\times10^{-8},\ \omega_{3,3}=1.026\times10^{-7},$$

and frequencies of $Gd_n(x_d,\tau)$:

$$\omega_{1,1}=1.370\times10^{-6},\ \omega_{2,1}=3.424\times10^{-7},\ \omega_{2,2}=1.223\times10^{-7},$$
$$\omega_{3,1}=1.522\times10^{-7},\ \omega_{3,2}=5.435\times10^{-8},\ \omega_{3,3}=7.609\times10^{-7},$$

were computed by iterations from tolerance equations

$$R_n(2K+1)\sum_{m=1}^{n}\omega_{n,m}\sum_{l=1}^{K}a_{2K,2l}\left(\left|A_{n,2l,m}\right|+\left|B_{n,2l,m}\right|\right)=2d_c,$$
$$R_n 2K\sum_{m=1}^{n}\omega_{n,m}\sum_{l=1}^{K}a_{2K-1,2l-1}\left(\left|A_{n,2l-1,m}\right|+\left|B_{n,2l-1,m}\right|\right)=2d_c. \tag{103}$$

These equations guarantee that absolute values of remainders of the structural approximations of (94)

$$rm_{n,2K}\left(x_d,\tau\right)=-R_n\left(2K+1\right)\sum_{m=1}^{n}\omega_{n,m}sh_{n,2K+2,m}\sum_{l=1}^{K}a_{2K,2l}\left(A_{n,2l,m}sn_{n,2l,m}+B_{n,2l,m}cs_{n,2l,m}\right), \tag{104}$$

$$rm_{n,2K-1}\left(x_d,\tau\right)=R_n\left(2K\right)\sum_{m=1}^{n}\omega_{n,m}sh_{n,2K+1,m}\sum_{l=1}^{K}a_{2K-1,2l-1}\left(A_{n,2l-1,m}sn_{n,2l-1,m}+B_{n,2l-1,m}cs_{n,2l-1,m}\right) \tag{105}$$

do not exceed $2d_c$ for all n, x_d, and τ, where $d_c=10^{-4}$ is an admissible error of computation of an invariant structure. Chaotic behavior of the boundary functions increases with n, K, and, especially, R_n, while various chaotic scenarios are modeled by $N\left[(N+1)(2K+1)+2\right]$ parameters R_n, Xa_n, $\omega_{n,m}$, $\tau_{n,m}$, $A_{n,2l,m}$, $B_{n,2l,m}$, $A_{n,2l-1,m}$, and $B_{n,2l-1,m}$.

The KEF structures and KEF-DEF structures are visualized in Figure 3 by instantaneous 3d surface plots with isocurves at $t=1.2\times10^6$ for scalar potential ϕ (60) and the dynamic pressure $p_d=g_z z+\left[p_t\left(x,z,t\right)-p_0\right]/\rho$ with p_t from (76), respectively. Local minimums of the KEF-DEF structure for p_d coincides with local maximums of the DEF structure for k_e, in agreement with [1] . The spatial structure of ϕ and p_d for stochastic waves coincides with that shown in Figure 2and Figure 3 of [5] for conservative

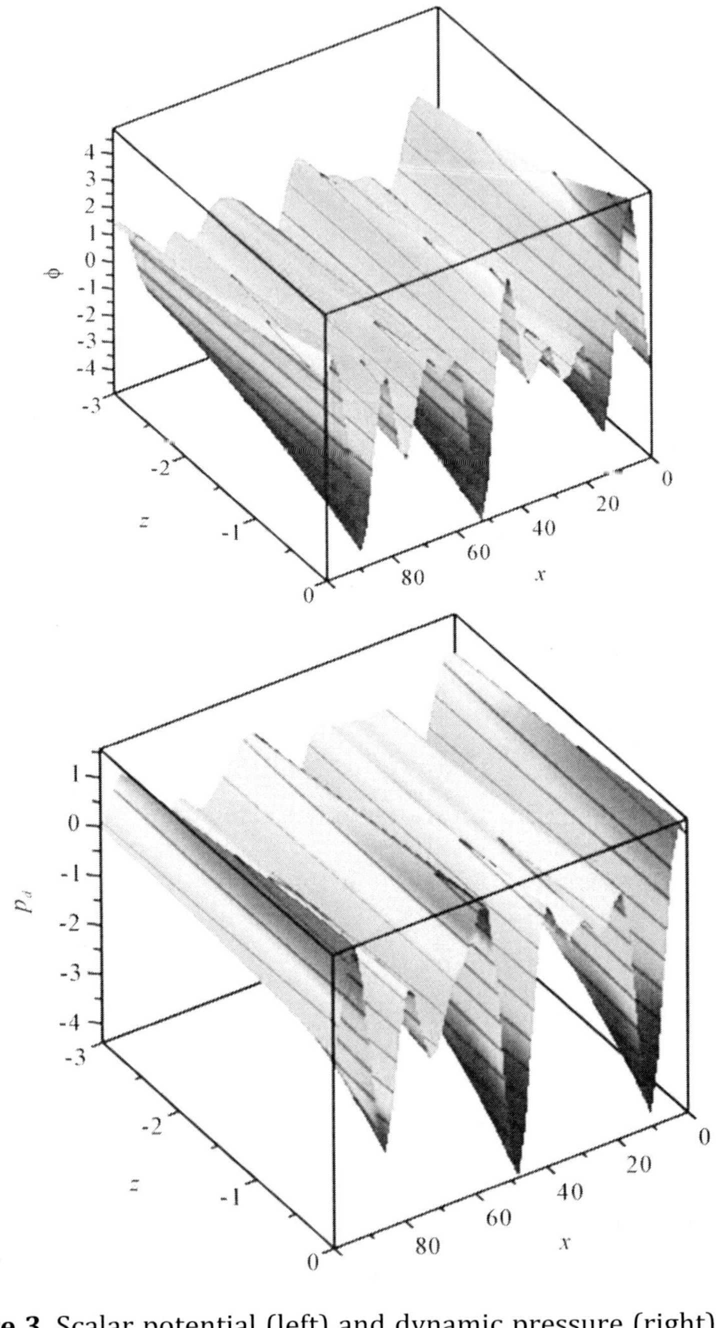

Figure 3. Scalar potential (left) and dynamic pressure (right) of the lower cumulative flow.

waves. The temporal structure of ϕ and p_d for stochastic waves, which is visualized by animations, significantly differs from that of ϕ and p_d for conservative waves because of the chaotic behavior of boundary functions $Fw_n(t)$ and $Gw_n(t)$.

5. CONCLUSIONS

The computational method of solving PDEs by decomposition in invariant structures, which continues the analytical methods of separation of variables and undetermined coefficients, is generalized in the current paper at the KEF, DEF, and KEF-DEF structures with time-dependent coefficients. This computational method is implemented in the kinematic and dynamic problems for internal waves by 43 procedures with 1070 code lines of the experimental computing in total and 35 procedures with 932 code lines of the theoretical computing in total. These structures with time-dependent structural coefficients are invariant with respect to various differential and algebraic operations.

For internal waves vanishing at infinity in the upper and lower domains, the Dirichlet problems are formulated for the kinematic and dynamic systems of the vorticity, continuity, Helmholtz, Lamb-Helmholtz, and Bernoulli equations. The exact solutions of the Navier-Stokes PDEs for the deterministic chaos of N stochastic waves are revealed experimentally, proved theoretically, and justified by the system of Navier-Stokes PDEs in the class of flows with the harmonic velocity field. The kinematic and dynamic solutions for stochastic waves coincide with the correspondent solutions for conservative waves [5] when stochastic boundary functions are reduced to constants.

Independence of both propagating and interacting internal waves is shown by computation of the Jacobian determinants in the DEF structures. Conditions for existence of parallel and orthogonal waves with time-de- pendent amplitudes are obtained through the Jacobian determinants, as well. The harmonic relationships between six pairs of the harmonic, fluid-dynamic variables, their temporal derivatives, and their spatial derivatives with respect to x and z are derived both for the upper and lower flows.

The stochastic boundary functions are constructed from the stochastic solutions of the one-dimensional Navier-Stokes equation [7] with hyperbolic temporal modes and trigonometric spatial modes in the HETO, HETE, HOTO, and HOTE structures. Various scenarios of a developed wave chaos are modeled by 3N parameters for internal waves and 2N stochastic boundary functions, which depend on $N[(N+1)(2K+1)+2]$ parameters, where K is a number of temporal modes.

ACKNOWLEDGEMENTS

Victor A. Miroshnikov The author thanks S. P. Bhavaraju for the stimulating discussion at the 2013 SIAM Annual Meeting. Support of the College of Mount Saint Vincent and CAAM is gratefully acknowledged.

REFERENCES

1. Lamb, S.H. (1945) Hydrodynamics. 6th Edition, Dover Publications, New York.

2. Pozrikidis, C. (2011) Introduction to Theoretical and Computational Fluid Dynamics. 2nd Edition, Oxford University Press, Oxford.

3. Miroshnikov, V.A. (2005) The Boussinesq-Rayleigh Series for Two-Dimensional Flows away from Boundaries. Applied Mathematics Research Express, 2005, 183-227.

4. Korn, G.A. and Korn, T.A. (2000) Mathematical Handbook for Scientists and Engineers: Definitions, Theorems, and Formulas for Reference and Review. 2nd Revised Edition, Dover Publications, New York.

5. Miroshnikov, V.A. (2014) Nonlinear Interaction of N Conservative Waves in Two Dimensions. American Journal of Computational Mathematics, 4, 127-142.

6. Kochin, N.E., Kibel, I.A. and Roze, N.V. (1964) Theoretical Hydromechanics. John Wiley & Sons Ltd., Chichester.

7. Miroshnikov, V.A. (2009) Spatiotemporal Cascades of Exposed and Hidden Perturbations of the Couette Flow. Advances and Applications in Fluid Dynamics, 6, 141-165.

8. Miroshnikov, V.A. (2012) Dual Perturbations of the Poiseuille-Hagen Flow in Invariant Elliptic Structures. Advances and Applications in Fluid Dynamics, 11, 1-58.

8

CHAPTER

Multivariate Stochastic Volatility Estimation with Sparse Grid Integration

Halil Erturk Esen

Computational Science and Engineering Department, Informatics Institute, Istanbul Technical University, Istanbul, Turkey

ABSTRACT

Multivariate stochastic volatility (MSV) models are nonlinear state space models that require either linear approximations or computationally demanding methods for handling the high dimensional integrals arising in the estimation problems of the latent volatilities and model parameters. Markov Chain Monte Carlo (MCMC) methods, which are based on Monte Carlo simulations using special sampling schemes, are by far the most studied method with several extensions and versions in previous stochastic volatility estimation studies. Exact nonlinear filters and particularly numerical integration based methods, such as the method proposed in this paper, were neglected and not studied as extensively as MCMC methods especially in the multivariate settings of stochastic volatility models. Filtering, smoothing, prediction and parameter estimation algorithms based on the sparse grid integration method are developed and proposed for a general MSV model. The proposed algorithms for estimation are compared with an implementation of MCMC based algorithms in a simulation study followed by an illustration of the proposed algorithms on empirical data of foreign exchange rate returns of US dollars and Euro. Results showed that the proposed algorithms based on the sparse grid integration method can be promising alternatives to the MCMC based algorithms especially in practical applications with their appealing characteristics.

Keywords: Multivariate Stochastic Volatility, Nonlinear State Space, Sparse Grid Integration, Markov Chain Monte Carlo

1. INTRODUCTION

Modeling, analysis, and estimation of volatilities of asset returns in financial markets have been a major research area in the last three decades because of the significant conceptual role of volatility in mathematical and quantitative finance. Reliable volatility estimates of asset returns are indispensible inputs to various mathematical models in financial frameworks including but not limited to risk management, option pricing, portfolio and asset management. A considerably rich literature on volatility research describes the well studied stylized facts about volatility including the time-varying nature of volatility, leverage effects, volatility spillovers, heavy tails and long memory. Comprehensive treatment of financial volatility and discussions on the stylized facts can be found in [1] -[4] .

Two main modeling approaches addressing the mentioned stylized facts are stochastic volatility (SV) and autoregressive conditionally heteroskedastic (ARCH) models. SV models differ from ARCH models by modeling the conditional volatility as a separate stochastic process having its own error term instead of explicitly modeling it. As discussed in [5] , this stochastic approach makes the SV models flexible and versatile in capturing the mentioned stylized facts; however these advantages come with computational costs associated with the estimation methods. SV models are essentially nonlinear state space models requiring approximations or computationally demanding numerical methods for the associated estimation problems.

Several extensions on the univariate SV models addressing the mentioned stylized facts have been studied and proposed after the SV model of Taylor which dealt only with volatility clustering as described in [6] [7] . After the first multivariate stochastic volatility (MSV) model in the literature, given in [8] , MSV modeling and estimation had become a rich research area composed of several model specifications addressing different and more complicated stylized facts, not only about volatilities but also about co-volatilities and their dynamics. [9] [10] , [11] are among many of those works and a review can be found in [12] . While addressing the stylized facts and flexibility, another objective was keeping the complexity under control in those MSV modeling efforts since dimensionality brought additional complexity on top of the inherent complexity due to the nonlinearity in SV models. Being non-linear state space models, even univariate SV models require methods that can handle high dimensional integrals in obtaining smoothing, filtering and prediction estimates of time-varying volatilities and parameter estimates as mentioned in [13] [14] . Furthermore, an extra complexity is introduced in MSV models due to the dimensionality of latent volatilities.

Several studies incorporated algorithms, providing either fast or simplified approximations based on the well- known Kalman filter and its extensions, using Laplace approximations, variations of moment matching and method of moments, and quasi likelihood methods. Although being fast and simple those methods generally suffered from poor performance as discussed in [2] [13] . Illustrations and examples of these methods can be found in [2] [4] [7] [8] [15] [16] .

Difficulties with estimation of SV models quickly incited the usage of computationally intensive simulation based Monte Carlo methods for better estimations with advances in computational resources. Several Monte Carlo based algorithms incorporating resampling, particle filters, rejection sampling and importance sampling algorithms have been proposed with examples that can be found in [13] [14] [17] [18] .

A major step in SV estimation was with studies such as [19] -[23] which introduced the Markov Chain Monte Carlo (MCMC) methods to the econometrics and SV fields. MCMC methods including the influential Metropolis-Hastings and Gibbs sampling algorithms quickly became central to the SV modeling and estimation studies, and an appreciable amount of literature on applications of different variations of MCMC methods was built up on several types of SV models, especially for the MSV models. Theoretically, MCMC methods are immune to the curse of dimensionality by construction, and they can easily be implemented in a Bayesian setting where the parameter estimation can also be handled without a maximization routine for the likelihood, hence without an explicit evaluation of the likelihood function as described in [23] . These appealing characteristics of MCMC methods made them a natural first choice in MSV estimation studies. However, MCMC algorithms are not flawless. They still require intense computational resources for complicated iterative sampling schemes for estimation. They are not exact as being Monte Carlo simulation based methods, and additional issues on error control and convergence are inherent particularly for MCMC methods.

Multidimensional integrals arising in estimation of SV models can be handled by classical numerical integration methods. Being exact methods with a deterministic nature, convergence properties of classical numerical integration methods are superior to probabilistic methods. However, when the problem dimension increases these methods become computationally infeasible since the number of dimensions increases the complexity of these type of algorithms exponentially. Unsurprisingly, studies on application of the numerical integration methods to nonlinear state space models and particularly to MSV models are quite rare compared to the approximation based methods and Monte Carlo simulation based methods including the MCMC. Refs. [14] and [24] are the main studies which discussed the application of numerical integration methods to nonlinear state space models.

Sparse grid integration method is a smartly reshaped version of classical numerical integration method to handle multidimensional integrals by constructing multi-dimensional integration formulas in a way that the dimensionality effect is decreased to a certain extent which allows practical implementation in high dimensional cases in contrast to the classical numeric integration methods. Sparse grid integration approach starts with [25] and detailed treatment of different approaches based on the idea is available in [26] [27] . The sparse grid integration approach was applied to some economics and financial problems (e.g., discrete choice analysis in [27] , collateral mortgage optimization problem in [28] , derivative and option pricing in [29] and asset liability in life insurance in [30]). However, estimation algorithms based on the sparse grid integration methods for SV models have been neither studied nor mentioned in the literature.

The goal of this paper is to show that the practical implementation of numerical integration method is possible by developing estimation algorithms incorporating the sparse grid integration methods as an alternative, and to depict that some important advantages of sparse grid integration methods over MCMC methods can be realized for MSV models. In this study, the sparse grid integration method is applied to the estimation problems of a given MSV model by developing density based estimation algorithms which include smoothing, filtering, prediction and parameter estimation incorporating the sparse grid integration method. Additionally, the estimation algorithms based on the proposed sparse grid integration method is illustrated on simulated and empirical data with comparisons.

The control of covariance matrices in MSV modeling and their handling in associated estimation algorithms are among the core topics of the MSV research. In this context, an approach which is new to the MSV modeling and estimation literature, to parameterize the correlation matrices, hence providing an alternative mechanism for handling covariance matrices in estimation, is also proposed and applied to the MSV model estimation algorithms in this paper. The approach is based on the slightly modified versions of the methods described in [31] [32] in a general setting.

The rest of this paper is organized as follows. A general multivariate stochastic volatility model and its state space representation are provided in Section 2. Development of estimation algorithms based on the proposed sparse grid integration method for the MSV model is presented and their properties are discussed in Section 3. An implementation of an MCMC algorithm for the MSV model is provided in Section 4. In Section 5, estimation algorithms with the proposed method are compared with the algorithms based on an MCMC implementation on simulated data. In Section 6, estimation algorithms with the proposed method are illustrated on empirical data. Finally, in Section 7 concluding remarks and further research directions are presented.

2. MULTIVARIATE STOCHASTIC VOLATILITY MODEL

The first MSV model in the literature is given in [8] . The specification of the basic MSV model based on this model starts with modeling the asset returns by

$$\boldsymbol{y}_t = \boldsymbol{V}_{y,t}^{1/2}\boldsymbol{\varepsilon}_t, \quad t = 1,\cdots,T, \tag{1}$$

where $\boldsymbol{y}_t = \left(y_{1t},\cdots,y_{pt}\right)'$ is a p-dimensional vector of asset returns, and $\boldsymbol{\varepsilon}_t = \left(\varepsilon_{1t},\cdots,\varepsilon_{pt}\right)'$ is a p-dimensional

vector of observation disturbances which have a zero mean and a correlation matrix, P_ε, defining the correlations between asset returns. In Equation (1), $\boldsymbol{V}_{y,t} = \mathrm{diag}\left(\exp\left(h_{1,t}\right),\cdots,\exp\left(h_{p,t}\right)\right)$ is a diagonal matrix whose diagonal entries are the time-varying variances and their logarithms follow an AR(1) process,

$$\boldsymbol{h}_{t+1} = \boldsymbol{\gamma} + \boldsymbol{\varphi}\boldsymbol{h}_t + \boldsymbol{\eta}_t, \quad t = 1,\cdots,T, \tag{2}$$

where, $\boldsymbol{h}_t = \left(h_{1t},\cdots,h_{pt}\right)'$ is the vector of unobserved log-volatilities, $\boldsymbol{\eta}_t = \left(\eta_{1t},\cdots,\eta_{pt}\right)'$ is a vector of distur-

bances with zero mean and a variance matrix $\boldsymbol{V}_\eta = \mathrm{diag}(\sigma_{1\eta}^2,..,\sigma_{p\eta}^2)$. In Equation (2), $\boldsymbol{\gamma} = \left(\gamma_1,\cdots,\gamma_p\right)'$ is the intercept parameter vector and $\boldsymbol{\varphi} = \mathrm{diag}(\varphi_1, \cdots, \varphi_p)$ is the diagonal matrix of persistence parameters of the AR(1) process which drives the time-varying log-volatilities.

In Equation (1), the error vector $\boldsymbol{\varepsilon}_t = (\varepsilon_{1t}, \cdots, \varepsilon_{pt})'$ enters the model multiplicatively, thus the model is nonlinear. Being a time-varying parameter model and nonlinear in structure, multivariate stochastic volatility model given by Equation (1) and Equation (2) is a non-linear state space model in essence. Equation (1) is considered to be the observation or measurement equation and implies the one step conditional density of observations (i.e. asset returns) given by

$$p_y(\boldsymbol{y}_t \mid \boldsymbol{h}_t, \boldsymbol{P}_\varepsilon), \quad t = 1, \cdots, T, \tag{3}$$

and Equation (2) is the state equation, and implies the one step conditional density of states (i.e., log-volatilities) given by

$$p_h(\boldsymbol{h}_t \mid \boldsymbol{h}_{t-1}, \boldsymbol{\gamma}, \boldsymbol{\varphi}, \boldsymbol{V}_\eta), \quad t = 1, \cdots, T, \tag{4}$$

where, the initial step has an assumed density, $p_h(\boldsymbol{h}_1 \mid \boldsymbol{h}_0, \boldsymbol{\gamma}, \boldsymbol{\varphi}, \boldsymbol{V}_\eta) \sim \mathrm{N}(\boldsymbol{\mu}_0, \boldsymbol{V}_0)$.

Several types of extensions and alternatives of the basic multivariate stochastic volatility model given in Equation (1) and Equation (2) have been proposed and implemented in the literature addressing various stylized facts about volatilities and co-volatilities. Examples can be found in [2] [4] [8] -[12] .

In this paper, the estimation algorithms with the proposed estimation algorithms and MCMC based estimation algorithms are implemented on the basic model given by Equation (1) and Equation (2) and this particular model will be referred as MSVb in the subsequent sections of this study.

3. ESTIMATION WITH SPARSE GRID INTEGRATION

3.1. Preliminaries

Similar to classical numerical integration methods, sparse grid integration methods are also based on integration formulas which are simply represented by a set of function evaluation points and corresponding weights. These points and weights are then used to evaluate the integral of a given function with

$$I^{(d)} f = \int_{\Omega} f(\boldsymbol{x}) \mathrm{d}\boldsymbol{x} \approx \sum_{i=1}^{N_l} \omega_{il} f(\boldsymbol{x}_{il}) , \quad (5)$$

where, d is the dimension, l is the level, x_{li} are the p-dimensional vectors representing points, $\omega_i$are the weights and N_l is the number of points of the integration formula which is represented by,

$$Q_l^{(d)} f = \sum_{i=1}^{N_l} \omega_{il} f(\boldsymbol{x}_{il}) . \quad (6)$$

In Equation (5) and Equation (6), level $l=1,2,\cdots$ determines the number of points, N_l, in the formula. Relationship between the level l and the number of points, N_l depends on the type of the formula. Further details on numeric integration and quadrature formulas can be found in [26] , and [27] .

In order to implement sparse grid integration to the MSVb model, sparse grid integration formula and the associated sparse grid is constructed first by the telescoping sum,

$$Q_l^{(p)} = \sum_{|\boldsymbol{k}|_1 \le l+p-1} \left(\Delta_{k_1}^{(1)} \otimes \cdots \otimes \Delta_{k_p}^{(1)} \right) , \quad (7)$$

where $Q_l^{(p)}$ is an integration formula of dimension p and level l, $\Delta_k^{(l)} = \left(Q_k^{(l)} - Q_{k-1}^{(l)} \right)$are the difference formulas obtained from a univariate integration formula having different levels, $\boldsymbol{k} = (k_1, \cdots, k_p)$ is a p-dimensional vector and $|\cdot|_r$ is the r-norm operator. Here,

Equation (7) simply constructs a p-dimensional integration formula from a given univariate integration formula of different levels. See [26] -[28] for details on the construction of regular sparse grids.

Classical construction of multivariate integration formulas differ in the approach used in Equation (7), that is, if $|k|_1 < l+p$, is replaced by $|k|_\infty < l$ in Equation (7), then the complete tensor grid used in classical approach would be obtained. The main difference is that, the regular sparse grid constructed by Equation (7) involves $O\left(N(\log N)^{p-1}\right)$ degrees of freedom with N points in one dimension of the grid at the boundary under certain conditions, while classical multivariate numerical integration formula has an exponential dependence, $O(N^p)$, on the number of dimensions. As a result, although not completely, sparse grid method significantly loosens the dependence on number of dimensions for practical application. For details on error bounds and complexity of sparse grids, see [26] -[28] .

Construction of the sparse grid starts with the selection of a univariate quadrature rule, $Q_l^{(1)}$, among many alternatives such as trapezoidal rule, rectangle rule, Simpson's rule, Clenshaw-Curtis and Gaussian quadrature rules including but not limited to Gauss-Hermite, Gauss-Legendre, and Gauss-Patterson. In this paper, the well- known iterated trapezoid rule for open interval is used:

$$Q_l^{(1)} f = \frac{1}{N_l+1}\left(\frac{3}{2} f\left(\frac{1}{N_l+1}\right) + \sum_{i=2}^{N_l-1} f\left(\frac{i}{N_l+1}\right) + \frac{3}{2} f\left(\frac{N_l}{N_{l+1}}\right)\right), \quad (8)$$

where the number of points for level l is $N_l = 2^l - 1$.

3.2. Smoothing, Filtering and Prediction Algorithms

In this section, the sparse grid method is applied to the density based estimation algorithms of the state-space model of MSVb. In the development of density based estimation algorithms for the MSVb model, a similar approach for the univariate nonlinear state-space models described in [14] and [24] is followed.

Let $Q_{l,t}^{(p)}$ be p-dimensional sparse grid quadrature rule constructed from the choice of univariate quadrature rule (i.e. trapezoid rule in our case), $Q_l^{(1)}$, of level l, at time t, and let $G_t = \{1, 2, \cdots, N_G\}$ be the set of sparse grid point indices where $N_l = 2^l - 1$ for the trapezoid rule given in Equation (8). Then, each point $i \in G_t$ in the grid is characterized with a p-dimensional vector $\boldsymbol{c}^i = \left(c_1^i, \cdots, c_p^i\right)'$ representing the grid coordinates of the points and corresponding weight ω_t^i. The constructed sparse grid can be applied to the state space model of MSVb by providing suitable integration intervals for each dimension of the state vector, $\boldsymbol{h}_t = \left(h_{1t}, \cdots, h_{pt}\right)$, for $t = 1, \cdots, T$. Once the integration intervals provided, grid coordinates of points, c^i, can be converted to actual point coordinates $\boldsymbol{h}_t^i$. A method for identifying suitable integration intervals for the states, h_t, is incorporating the estimates form an extended Kalman filter applied to the MSVb model and setting integration intervals as $\left(\hat{\boldsymbol{h}}_{t|s} - v\hat{\boldsymbol{d}}_{t|s}, \hat{\boldsymbol{h}} + v\hat{\boldsymbol{d}}_{t|s}\right)$, where $\hat{\boldsymbol{h}}_{t|s}$ is the state estimates and $\hat{\boldsymbol{d}}_{t|s}$ is the vector composed of square roots of the diagonal elements of the covariance estimate, $\hat{\boldsymbol{\Sigma}}_{t|s}$ (i.e., standard deviation estimates of states) obtained with the extended Kalman filter, and v is a scalar. In this approach, appropriate Kalman filter algorithm is selected depending on the type of the estimation problem, namely filtering (s = t), smoothing (s = T), or prediction (s = T + L). See [4] [8] [14] [15] for different usages of the Kalman filter in SV estimation problems, and [4] and [14] for details and derivation of the Kalman filter estimation algorithms.

Conditional filtering density of states for the MSVb model can be obtained recursively by the following two-step algorithm:

$$p\left(\boldsymbol{h}_t \mid \boldsymbol{Y}_{t-1}, \boldsymbol{\gamma}, \boldsymbol{\varphi}, \boldsymbol{V}_\eta, \boldsymbol{P}_\varepsilon\right) = \int p_h\left(\boldsymbol{h}_t \mid \boldsymbol{h}_{t-1}, \boldsymbol{\gamma}, \boldsymbol{\varphi}, \boldsymbol{V}_\eta\right) p\left(\boldsymbol{h}_{t-1} \mid \boldsymbol{Y}_{t-1}, \boldsymbol{\gamma}, \boldsymbol{\varphi}, \boldsymbol{V}_\eta, \boldsymbol{P}_\varepsilon\right) \mathrm{d}\boldsymbol{h}_{t-1},$$

$$p\left(\boldsymbol{h}_t \mid \boldsymbol{Y}_t, \boldsymbol{\gamma}, \boldsymbol{\varphi}, \boldsymbol{V}_\eta\right) = \frac{p_y\left(\boldsymbol{y}_t \mid \boldsymbol{h}_t, \boldsymbol{P}_\varepsilon\right) p\left(\boldsymbol{h}_t \mid \boldsymbol{Y}_{t-1}, \boldsymbol{\gamma}, \boldsymbol{\varphi}, \boldsymbol{V}_\eta, \boldsymbol{P}_\varepsilon\right)}{\int p_y\left(\boldsymbol{y}_t \mid \boldsymbol{h}_t, \boldsymbol{P}_\varepsilon\right) p\left(\boldsymbol{h}_t \mid \boldsymbol{Y}_{t-1}, \boldsymbol{\gamma}, \boldsymbol{\varphi}, \boldsymbol{V}_\eta, \boldsymbol{P}_\varepsilon\right) \mathrm{d}\boldsymbol{h}_t}, \quad (9)$$

for $t=1,\cdots,T$. Equipped with the sparse grid points, weights, and set of their indices, the integrals in the filtering algorithm can be handled numerically as

$$p\left(\boldsymbol{h}_t^i \mid \boldsymbol{Y}_{t-1},\gamma,\varphi,\boldsymbol{V}_\eta,\boldsymbol{P}_\varepsilon\right)\approx \sum_{j\in G_{t-1}} p_h\left(\boldsymbol{h}_t^i \mid \boldsymbol{h}_{t-1}^j,\gamma,\varphi,\boldsymbol{V}_\eta\right)p\left(\boldsymbol{h}_{t-1}^j \mid \boldsymbol{Y}_{t-1},\gamma,\varphi,\boldsymbol{V}_\eta,\boldsymbol{P}_\varepsilon\right)\omega_{t-1}^j,\quad i\in G_t,$$

$$p\left(\boldsymbol{h}_t^i \mid \boldsymbol{Y}_t,\gamma,\varphi,\boldsymbol{V}_\eta\right)=\frac{p_y\left(\boldsymbol{y}_t \mid \boldsymbol{h}_t^i,\boldsymbol{P}_\varepsilon\right)p\left(\boldsymbol{h}_t^i \mid \boldsymbol{Y}_{t-1},\gamma,\varphi,\boldsymbol{V}_\eta,\boldsymbol{P}_\varepsilon\right)}{\sum_{i\in G_t} p_y\left(\boldsymbol{y}_t \mid \boldsymbol{h}_t^i,\boldsymbol{P}_\varepsilon\right)p\left(\boldsymbol{h}_t^i \mid \boldsymbol{Y}_{t-1},\gamma,\varphi,\boldsymbol{V}_\eta,\boldsymbol{P}_\varepsilon\right)\omega_t^i}.$$

(10)

Starting with the filtering density obtained in Equation (10), L-step ahead prediction density can be recursively obtained by

$$p\left(\boldsymbol{h}_{t+L} \mid \boldsymbol{Y}_t,\gamma,\varphi,\boldsymbol{V}_\eta,\boldsymbol{P}_\varepsilon\right)=\int p_h\left(\boldsymbol{h}_{t+L} \mid \boldsymbol{h}_{t+L-1},\gamma,\varphi,\boldsymbol{V}_\eta\right)p\left(\boldsymbol{h}_{t+L-1} \mid \boldsymbol{Y}_t,\gamma,\varphi,\boldsymbol{V}_\eta,\boldsymbol{P}_\varepsilon\right)\mathrm{d}\boldsymbol{h}_{t+L-1},$$

(11)

and using the sparse grid integration, approximated by

$$p\left(\boldsymbol{h}_{t+L}^i \mid \boldsymbol{Y}_t,\gamma,\varphi,\boldsymbol{V}_\eta,\boldsymbol{P}_\varepsilon\right)=\sum_{j\in G_{t+L-1}} p_h\left(\boldsymbol{h}_{t+L}^i \mid \boldsymbol{h}_{t+L-1}^j,\gamma,\varphi,\boldsymbol{V}_\eta\right)p\left(\boldsymbol{h}_{t+L-1}^j \mid \boldsymbol{Y}_t,\gamma,\varphi,\boldsymbol{V}_\eta,\boldsymbol{P}_\varepsilon\right)\omega_{t+L-1}^j,\quad i\in G_{t+L}.$$

(12)

Given the filtering density, smoothing density is obtained with the backward recursion,

$$p\left(\boldsymbol{h}_t \mid \boldsymbol{Y}_T,\gamma,\varphi,\boldsymbol{V}_\eta,\boldsymbol{P}_\varepsilon\right)=p\left(\boldsymbol{h}_t \mid \boldsymbol{Y}_t,\gamma,\varphi,\boldsymbol{V}_\eta,\boldsymbol{P}_\varepsilon\right)\int\frac{p\left(\boldsymbol{h}_{t+1} \mid \boldsymbol{Y}_T,\gamma,\varphi,\boldsymbol{V}_\eta,\boldsymbol{P}_\varepsilon\right)p_h\left(\boldsymbol{h}_{t+1} \mid \boldsymbol{h}_t,\gamma,\varphi,\boldsymbol{V}_\eta\right)}{p\left(\boldsymbol{h}_{t+1} \mid \boldsymbol{Y}_t,\gamma,\varphi,\boldsymbol{V}_\eta,\boldsymbol{P}_\varepsilon\right)}\mathrm{d}\boldsymbol{h}_{t+1},$$

(13)

and sparse grid integration approximation is given by

$$p\left(\boldsymbol{h}_t^i \mid \boldsymbol{Y}_T, \gamma, \varphi, \boldsymbol{V}_\eta, \boldsymbol{P}_\varepsilon\right)$$
$$= p\left(\boldsymbol{h}_t^i \mid \boldsymbol{Y}_t, \gamma, \varphi, \boldsymbol{V}_\eta, \boldsymbol{P}_\varepsilon\right) \sum_{i \in G_t} \frac{p\left(\boldsymbol{h}_{t+1}^j \mid \boldsymbol{Y}_T, \gamma, \varphi, \boldsymbol{V}_\eta, \boldsymbol{P}_\varepsilon\right) p_h\left(\boldsymbol{h}_{t+1}^j \mid \boldsymbol{h}_t^i, \gamma, \varphi, \boldsymbol{V}_\eta\right)}{p\left(\boldsymbol{h}_{t+1}^j \mid \boldsymbol{Y}_t, \gamma, \varphi, \boldsymbol{V}_\eta, \boldsymbol{P}_\varepsilon\right)} \omega_t^i, \quad j \in G_{t+1}, \tag{14}$$

for $t = T-1, T-2, \cdots, 1$.

Once the estimation density of interest is obtained with one of the recursive algorithms approximated by sparse grid integration, estimation of statistical properties such as the mean and the variance of a function, $f(\boldsymbol{h}_r)$ can be computed by the following expectation and its approximation with sparse grid integration:

$$E\left[f(\boldsymbol{h}_r) \mid \boldsymbol{Y}_s\right] = \int f(\boldsymbol{h}_r) p\left(h_r \mid \boldsymbol{Y}_s, \gamma, \varphi, \boldsymbol{V}_\eta\right) \mathrm{d}\boldsymbol{h}_i \approx \sum_{i \in G_r} f\left(\boldsymbol{h}_r^i\right) p\left(\boldsymbol{h}_r^i \mid \boldsymbol{Y}_s, \gamma, \varphi, \boldsymbol{V}_\eta\right) \omega_r^i, \tag{15}$$

where, $(r,s) = (T,T), (t,t), (t+L,t)$ for smoothing, filtering and prediction respectively.

Estimation algorithms derived so far, with the exception of smoothing, are all sequential (i.e., on-line) algorithms which do not require all the past information at each time period; instead the estimations from the previous step are used. This is important and advantageous from the computational perspective especially in practical applications, because the computational burden in prediction and filtering can be split across time periods in contrast with batch-type algorithms.

A possible issue with the recursive algorithms with sparse grid integration is the accumulation of numerical errors during recursions, where controls and corrections satisfying the requirement,

$$\int p\left(\boldsymbol{h}_r \mid \boldsymbol{Y}_s\right) \mathrm{d}\boldsymbol{h}_r = \sum_{i \in G_r} p\left(\boldsymbol{h}_r^i \mid \boldsymbol{Y}_s\right) \omega_r^i = 1 \tag{16}$$

can be implemented easily with only an extra computational cost of recalculation of $p(h_t \mid Y_t)$ as needed.

An advantage of the algorithms presented in this section based on the sparse grid integration is their computational parallelizability on multi-core and/or massively parallel architectures such as graphics processing units (GPUs). The recursive filtering algorithm given in Equation (10) can be effectively parallelized at each time period by assigning sparse grid nodes and corresponding computations to parallel processors with a simple load balancing strategy since they are not dependent to each other. This provides significant speed-up in implementation. Since the prediction and smoothing algorithms use the filtering algorithm, same speed-ups are effective for those as well in a parallel implementation.

3.3. Parameter Estimation

Parameter estimation for the MSVb model can be performed by maximizing the log-likelihood function given by

$$\begin{aligned}\log\left(p\left(Y_T \mid \gamma,\varphi,V_\eta,P_\varepsilon\right)\right) &= \log\left(\prod_{t=1}^{T}\left(\int p_y\left(y_t \mid h_t,P_\varepsilon\right)p\left(h_t \mid Y_{t-1},\gamma,\varphi,V_\eta,P_\varepsilon\right)\mathrm{d}h_t\right)\right)\\ &= \sum_{t=1}^{T}\log\left(\int p_y\left(y_t \mid h_t,P_\varepsilon\right)p\left(h_t \mid Y_{t-1},\gamma,\varphi,V_\eta,P_\varepsilon\right)\mathrm{d}h_t\right), \end{aligned} \tag{17}$$

and approximated with sparse grid integration by

$$\log\left(p\left(Y_T \mid \gamma,\varphi,V_\eta,P_\varepsilon\right)\right) = \sum_{t=1}^{T}\log\left(\sum_{i\in G_t} p_y\left(y_t \mid h_t^i,P_\varepsilon\right)p\left(h_t^i \mid Y_{t-1},\gamma,\varphi,V_\eta,P_\varepsilon\right)\omega_t^i\right), \tag{18}$$

with respect to the parameters γ, φ, V_η, and P_ε. Here, the log-likelihood function is actually the denominator of the update step of filtering algorithm given in Equation (10) and readily available as a byproduct of filtering algorithm. Therefore, obtaining the log-likelihood is also sequential as discussed in the Section 3.2 and no extra computational burden is incurred. Parallelizability advantage described in Section 3.2 is also effective for parameter estimation as it is based on filtering algorithm.

A challenge that must be sorted out in practical implementation of maximization routine for the parameter estimation is finding a way to handle the correlation matrix P_ε. Correlation matrix, P_ε, is a symmetric matrix with ones in the diagonal, off-diagonal entries having values in the interval (−1,1), and moreover it must be positive-definite. To handle the mentioned restrictions on the correlation matrix, P_ε and to be able to incorporate common non-linear optimization methods and tools in the parameter estimation problem, a parameterization approach of the correlation matrix, P_ε, which is a slightly different version of the parameterization described in [31] -[33] in a general setting, was proposed in this study.

In the proposed parameterization approach, positive definiteness of the correlation matrix is ensured by the relation, $P_\varepsilon = BB'$ where the (p × p) matrix B is

$$B = \begin{bmatrix} b_{1,1} & b_{1,2} & & b_{1,p} \\ b_{2,1} & & & 0 \\ & & 0 & 0 \\ b_{p,1} & 0 & 0 & 0 \end{bmatrix} \tag{19}$$

and its entries are constructed with the relation,

$$b_{ij} = \begin{cases} \cos(\alpha_{ij}) \prod_{k=1}^{j-1} \sin(\alpha_{ik}), & \text{if } j \geq i \\ 0, & \text{otherwise,} \end{cases} \tag{20}$$

where $i = 1, \cdots, p,\ j = 1, \cdots, p$ and $0 \leq \alpha_{ij} \leq \pi$. Let angles, α_{ij}, be the entries of a (p × p) matrix,

$$A = \begin{bmatrix} \alpha_{11} & & \alpha_{p-1,1} & 0 \\ & & 0 & 0 \\ \alpha_{1,p-1} & 0 & 0 & 0 \\ 0 & 0 & 0 & 0 \end{bmatrix}. \qquad (21)$$

In matrix A, there are $p\times(p-1)/2$ angles for a given (p × p) correlation matrix, P_ε. The parameterization given in Equations (19)-(21) allow one to replace the parameter matrix, P_ε, with matrix A in the maximization routine since matrix A does not have complex restriction on its entries for common optimization methods. In the notations of the estimation algorithms developed so far, P_ε can be replaced by A.

The log-likelihood function obtained by Equation (18) can be maximized using a general purpose nonlinear optimization method. Well known Newton's method, derivative free search algorithms such as Nelder-Mead, or quasi-Newton methods such as Broyden-Fletcher-Goldfarb-Shanno (BFGS) are some of the examples. Using derivative free optimization methods can significantly improve computational performance. However, methods calculating or approximating the Hessian matrix, H, at the maximum log-likelihood readily provide the observed Fisher information matrix, I in accordance with

$$\boldsymbol{I}(\tau) = -\boldsymbol{H}(\tau) = \frac{\partial^2}{\partial\tau_i \partial\tau_j} l(\tau), \qquad (22)$$

where τ is the vector of parameters where the elements of γ, φ, V_η, and A are stacked in order. Then, the standard errors for parameter estimates can be obtained easily from the inverse of the observed Fisher information matrix at the maximum log-likelihood, $I^{-1}(\tau_{ML})$.

4. ESTIMATION WITH MARKOV CHAIN MONTE CARLO

MCMC methods are simulation based methods that are proved to be quite successful and that are probably the most preferred methods in SV modeling and estimation. There is a large collection of papers with extensions and variations of MCMC methods for estimating SV models (e.g., [2] [9] -[11]).

MCMC methods directly sample from the conditional joint density of interest, which is often the conditional joint smoothing density of states, through dependent samples obtained with the help of a constructed Markov Chain so that the limiting probability is the conditional estimation density of interest.

For parameter estimation, two main approaches are available. First approach incorporates maximization of the log-likelihood in an expectation maximization (EM) algorithm or approximations with other Monte Carlo methods that use smoothing estimates obtained by the MCMC as described in [2] and [14] . Second approach is augmenting the parameter space to the state space and obtaining the estimations of parameters along with the smoothing estimates of states through MCMC sampling in a Bayesian setting without the need for explicitly calculating or approximating the log-likelihood and a maximization routine as described in [9] -[11] [23] . In this paper, the former approach is followed for implementation, that is, a MCMC algorithm which samples from the conditional joint smoothing density of states is implemented first and then an EM algorithm is implemented using the samples obtained by the MCMC.

Let, $\boldsymbol{H}_t = \{\boldsymbol{h}_1, \boldsymbol{h}_2, \cdots, \boldsymbol{h}_t\}$, be the set of state vectors and $\boldsymbol{Y}_t = \{\boldsymbol{y}_1, \boldsymbol{y}_2, \cdots, \boldsymbol{y}_t\}$ be the set of observation vectors up to time t, where $t = 1, \cdots, T$.
If $\boldsymbol{H}_{\setminus t} = \{\boldsymbol{h}_1, \cdots, \boldsymbol{h}_{t-1}, \boldsymbol{h}_{t+1}, \cdots, \boldsymbol{h}_t\}$ is defined as the set of state vectors up to time T excluding the state vector of time t, then sampling from the conditional joint smoothing density $p(\boldsymbol{H}_T \mid \boldsymbol{Y}_T, \boldsymbol{\gamma}, \boldsymbol{\varphi}, \boldsymbol{V}_\eta, \boldsymbol{P}_\varepsilon)$ can be performed with Gibbs sampling at each step sampling from the kernel density

$p\left(\boldsymbol{h}_t \mid \boldsymbol{H}_{\setminus t}, \boldsymbol{Y}_T, \gamma, \boldsymbol{\varphi}, \boldsymbol{V}_\eta, \boldsymbol{P}_\varepsilon\right)$ where,

$$p\left(\boldsymbol{h}_t \mid \boldsymbol{H}_{\setminus t}, \boldsymbol{Y}_T, \gamma, \boldsymbol{\varphi}, \boldsymbol{V}_\eta, \boldsymbol{P}_\varepsilon\right)$$
$$\propto \begin{cases} p_h\left(\boldsymbol{h}_t \mid \boldsymbol{h}_{t-1}, \gamma, \boldsymbol{\varphi}, \boldsymbol{V}_\eta\right) p_h\left(\boldsymbol{h}_{t+1} \mid \boldsymbol{h}_t, \gamma, \boldsymbol{\varphi}, \boldsymbol{V}_\eta\right) p_y\left(\boldsymbol{y}_t \mid \boldsymbol{h}_t, \boldsymbol{P}_\varepsilon\right), & \text{if } t = 1, \cdots, T-1, \\ p_h\left(\boldsymbol{h}_t \mid \boldsymbol{h}_{t-1}, \gamma, \boldsymbol{\varphi}, \boldsymbol{V}_\eta\right) p_y\left(\boldsymbol{y}_t \mid \boldsymbol{h}_t, \boldsymbol{P}_\varepsilon\right), & \text{if } t = T. \end{cases} \tag{23}$$

Given the observations and parameters, sampling from this kernel N times for $t = 1, \cdots, T$ completes the MCMC algorithm and the joint smoothing density $p\left(\boldsymbol{H}_t \mid \boldsymbol{Y}_T, \gamma, \boldsymbol{\varphi}, \boldsymbol{V}_\eta, \boldsymbol{P}_\varepsilon\right)$ is obtained. Usually, the first M samples are discarded as burn-in samples depending on the mixing and convergence properties of the chain.

The kernel given in Equation (23) does not allow direct sampling since it is not in an analytically tractable form, thus Metropolis-Hastings algorithm with a choice of proposal density, $p_*(\boldsymbol{h}_t)$ and Metropolis-Hasting weight,

$$w_{MH} = \begin{cases} \dfrac{p_h\left(\boldsymbol{h}_t^* \mid \boldsymbol{h}_{t-1}, \gamma, \boldsymbol{\varphi}, \boldsymbol{V}_\eta\right) p_h\left(\boldsymbol{h}_{t+1}^* \mid \boldsymbol{h}_t, \gamma, \boldsymbol{\varphi}, \boldsymbol{V}_\eta\right) p_y\left(\boldsymbol{y}_t \mid \boldsymbol{h}_t^*, \boldsymbol{P}_\varepsilon\right) p_*\left(\boldsymbol{h}_t\right)}{p_h\left(\boldsymbol{h}_t \mid \boldsymbol{h}_{t-1}, \gamma, \boldsymbol{\varphi}, \boldsymbol{V}_\eta\right) p_h\left(\boldsymbol{h}_{t+1} \mid \boldsymbol{h}_t, \gamma, \boldsymbol{\varphi}, \boldsymbol{V}_\eta\right) p_y\left(\boldsymbol{y}_t \mid \boldsymbol{h}_t, \boldsymbol{P}_\varepsilon\right) p_*\left(\boldsymbol{h}_t^*\right)}, & \text{if } t = 1, \cdots, T-1 \\ \dfrac{p_h\left(\boldsymbol{h}_t^* \mid \boldsymbol{h}_{t-1}, \gamma, \boldsymbol{\varphi}, \boldsymbol{V}_\eta\right) p_y\left(\boldsymbol{y}_t \mid \boldsymbol{h}_t^*, \boldsymbol{P}_\varepsilon\right) p_*\left(\boldsymbol{h}_t\right)}{p_h\left(\boldsymbol{h}_t \mid \boldsymbol{h}_{t-1}, \gamma, \boldsymbol{\varphi}, \boldsymbol{V}_\eta\right) p_y\left(\boldsymbol{y}_t \mid \boldsymbol{h}_t, \boldsymbol{P}_\varepsilon\right) p_*\left(\boldsymbol{h}_t^*\right)}, & \text{if } t = T, \end{cases} \tag{24}$$

can be implemented for sampling. In the Metropolis-Hastings step, a candidate, generated from the proposal density, $p_*(\boldsymbol{h}_t)$, is accepted with probability $\min(w_{MH}, 1)$ otherwise previous sample is kept as current sample. As a proposal density, several alternatives are available with examples in [14] . Following the same approach described in [14] , a normal density based on the extended Kalman filter,

$$p_{*}\left(\boldsymbol{h}_{t|T}\right)=f_{N}\left(\boldsymbol{h}_{t};\hat{\boldsymbol{h}}_{t|T},v\hat{\boldsymbol{\Sigma}}_{t|T}\right), \tag{25}$$

can be incorporated where $\hat{\boldsymbol{h}}_t$ and $\hat{\boldsymbol{\Sigma}}_t$ are the extended Kalman filter smoothing estimates of states and covariance matrices respectively and v is a scalar.

Filtering density is simply obtained by repeating the above procedure for smoothing T times, at each step replacing T by t. Once the filtering density at time T or equivalently smoothing density at time T is obtained, prediction density can be constructed through resampling with the following recursive procedure:

$$p\left(\boldsymbol{h}_{T+L}\mid \boldsymbol{Y}_T,\gamma,\varphi,\boldsymbol{V}_\eta,\boldsymbol{P}_\varepsilon\right)\approx\frac{1}{N}\sum_{j=1}^{N}p\left(\boldsymbol{h}_{T+L}\mid \boldsymbol{h}_{j,T+L-1},\gamma,\varphi,\boldsymbol{V}_\eta\right). \tag{26}$$

Parameter estimation incorporates the joint smoothing density of states given the observations and parameters, $p\left(\boldsymbol{H}_T\mid \boldsymbol{Y}_T,\gamma,\varphi,\boldsymbol{V}_\eta,\boldsymbol{P}_\varepsilon\right)$, obtained with the smoothing procedure above, in an expectation maximization algorithm, where at each iteration samples from, $p\left(\boldsymbol{H}_T\mid \boldsymbol{Y}_T,\gamma,\varphi,\boldsymbol{V}_\eta,\boldsymbol{P}_\varepsilon\right)$ are drawn and the expected log-likelihood function,

$$\begin{aligned}
&\mathrm{E}_{H|Y,\gamma,\varphi,V_\eta,P_\varepsilon}\left[\log\left(p\left(\boldsymbol{H}_T,\boldsymbol{Y}_T\mid\gamma,\varphi,\boldsymbol{V}_\eta,\boldsymbol{P}_\varepsilon\right)\right)\right]\\
&=\mathrm{E}_{H|Y,\gamma,\varphi,V_\eta,P_\varepsilon}\left[\log\left(p_h\left(\boldsymbol{H}_T\mid\gamma,\varphi,\boldsymbol{V}_\eta\right)p_y\left(\boldsymbol{Y}_T\mid\boldsymbol{H}_T,\boldsymbol{P}_\varepsilon\right)\right)\right]\\
&\approx\sum_{i=M+1}^{N}\log\left(p_h\left(\boldsymbol{H}_T^i\mid\gamma,\varphi,\boldsymbol{V}_\eta\right)p_y\left(\boldsymbol{Y}_T\mid\boldsymbol{H}_T^i,\boldsymbol{P}_\varepsilon\right)\right),
\end{aligned} \tag{27}$$

is calculated from those samples and maximized with respect to the parameters γ, φ, V_η, and $P_\varepsilon$followed by the next iteration starting with sampling again from $p\left(\boldsymbol{H}_T\mid \boldsymbol{Y}_T,\gamma,\varphi,\boldsymbol{V}_\eta,\boldsymbol{P}_\varepsilon\right)$ with the explained

MCMC procedure for smoothing using the updated values of parameters.

In the implementation of maximization routine, the same approach for parameterization of the correlation matrix, P_ε, described in Equations (19)-(21) in Section 3.3 is used to ensure the restrictions on the correlation matrix.

Filtering and prediction are often hard tasks for MCMC methods, while the real advantage of MCMC is apparent in smoothing and parameter estimation problems because the filtering, prediction and parameter estimation with the above MCMC based estimation algorithms require construction of the smoothing density first, in contrast with the estimation approaches given in the Section 3.2. Furthermore, the observations, y_t, for all time periods $t=1,\cdots,T$ are required to construct the smoothing density, hence for the filtering and prediction densities and parameter estimations too. Therefore, MCMC based estimation algorithms are called batch algorithms, whereas the estimation algorithms given in Section 3.2 were sequential (i.e. on-line) algorithms. Batch algorithms use all the information from the beginning to the last period, and all information are used again with the arrival of new information. Additionally, computational burden cannot be split across time periods which is a clear disadvantage for practical applications.

From the computational point of view, although being advantageous in serial implementations, MCMC based algorithms are generally not suitable for parallelization since they are batch algorithms producing dependent samples at each iteration where both temporal and spatial dependence exists. A limited parallelization is still possible in the implementation approach used in this paper though. Furthermore, parallelization in the implementations, which conduct parameter estimation through MCMC sampling in a Bayesian setting, is much harder.

5. APPLICATION ON SIMULATED DATA

This section illustrates the implementation of the proposed sparse grid integration method and compares the method with the implemented MCMC method on simulated data. The data are simulated using the MSVb model specifications with the following parameters:

$$\gamma = \begin{pmatrix} -0.25 \\ -0.25 \\ -0.25 \end{pmatrix}, \quad \varphi = \begin{pmatrix} 0.95 \\ 0.95 \\ 0.95 \end{pmatrix}, \quad V_\eta = \begin{pmatrix} 0.04 & 0 & 0 \\ 0 & 0.04 & 0 \\ 0 & 0 & 0.04 \end{pmatrix}, \quad P_\varepsilon = \begin{pmatrix} 1 & 0.6 & 0.6 \\ 0.6 & 1 & 0.6 \\ 0.6 & 0.6 & 1 \end{pmatrix},$$

which are close to typical values in previous empirical studies in the stochastic volatility literature. Using these parameters, observations y_t, and log-volatilities, h_t, are generated for T = 1,000 periods. The proposed sparse grid integration method and the MCMC method described in the previous sections are applied on the simulated data to find filtering, smoothing one-step prediction estimates of log-volatilities, and parameter estimates. This procedure is repeated R = 100 times to capture the statistics on estimations. The root mean squared error (RMSE) of the log-volatility estimates is calculated by

$$\text{RMSE-h}_{t|s} = \frac{1}{T}\sum_{t=t_S}^{T}\sqrt{\frac{1}{R}\sum_{r=1}^{R}\left(h_{t|s}^{(r)} - h_t^{(r)}\right)^2} \tag{28}$$

where $h_{t|s}^{(r)}$ is the log-volatility estimate obtained with the estimation method, and $h_t^{(r)}$ is the log-volatility from the simulated data where (s, t_S) = (t, 1) for filtering, (s, t_S) = (T, 1) for smoothing and (s, t_S) = (2, t – 1) for prediction. Similarly, the RMSE of the parameters are calculated by

$$\text{RMSE-}\tau = \sqrt{\frac{1}{R}\sum_{r=1}^{R}\left(\hat{\tau}^{(r)} - \tau^{(r)}\right)^2} \tag{29}$$

where $\hat{\tau}^{(r)}$ is the parameter estimated with the estimation method and $\tau_t^{(r)}$ is the actual values of the parameters which are used in generating the simulated data.

Using the MCMC algorithm described in Section 4, N = 1,000,000 samples are generated and M = 200,000 samples are discarded as burn-in samples. In the sparse grid integration method, univariate trapezoid rule given in Equation (8) is implemented with level $l = 7$ which leads to 127 points in each dimension at the boundary and a total of 2815 points in the three-dimensional sparse grid at each time period.

The scalar used in identification of the integration intervals and the proposal density in the MCMC algorithm with the Kalman filter covariance estimates was taken as $v = 6.0$. For parameter estimation, Newton's method is used as the maximization routine in both methods.

To speed-up the computations, graphical processing unit (GPU) computing is used in the implementations of the sparse grid integration method algorithms and this provided a significant computational advantage since the estimation algorithms described in Section 3.2 are highly parallelizable.

Table 1 shows the estimation results for the parameters, where the mean and the RMSE of parameter estimates are calculated from the simulation repeats using the formula in Equation (29). In general, parameter estimates are quite close in both methods when the means and true values are compared. However, the proposed sparse grid integration performs slightly better when RMSE values are considered with lower values for most of the parameters with exceptions γ_1 and σ_1. The level of the integration formula used, $l = 7$, provided enough accuracy that can be compared with that of MCMC method with 1,000,000 samples in this simulation study. However, for different choices of parameters especially for higher persistence parameter values the minimum requirement for accuracy levels would change.

Table 2 shows the RMSE of the estimated log-volatilities calculated by Equation (28) for comparison of latent log-volatility estimates generated by the two methods for filtering, smoothing and prediction problems. It can be observed that the RMSEs for all estimation problems are close with slightly better performance of sparse grid integration method in filtering and one-step prediction estimations while a better performance of MCMC method is observed in the smoothing estimates inTable 2.

Table 1. Estimation results for parameters.

	i	True	Sparse grid integration		MCMC	
			Mean	RMSE	Mean	RMSE
γ_i	1	−0.25	−0.237	0.072	−0.260	0.067
	2	−0.25	−0.248	0.029	−0.261	0.071
	3	−0.25	−0.245	0.034	−0.231	0.096
φ_i	1	0.95	0.947	0.024	0.956	0.069
	2	0.95	0.952	0.016	0.959	0.065
	3	0.95	0.946	0.026	0.961	0.071
$\sigma_{\eta i}$	1	0.2	0.209	0.045	0.206	0.040
	2	0.2	0.191	0.042	0.207	0.054
	3	0.2	0.205	0.037	0.214	0.067
ρ_{ij}	12	0.6	0.605	0.053	0.592	0.059
	13	0.6	0.594	0.041	0.590	0.073
	23	0.6	0.606	0.039	0.592	0.058

Table 2. Root mean squared errors for estimated log-volatilities, h_t.

	i	RMSEs for sparse grid integration			RMSEs for MCMC		
		Filtering, $h_{t\|t}$	Smoothing, $h_{t\|T}$	Prediction, $h_{t+1\|t}$	Filtering, $h_{t\|t}$	Smoothing, $h_{t\|T}$	Prediction, $h_{t+1\|t}$
h_i	1	0.204	0.195	0.246	0.208	0.185	0.251
	2	0.196	0.186	0.238	0.199	0.179	0.245
	3	0.199	0.189	0.233	0.203	0.182	0.241

6. EMPIRICAL APPLICATION

In this section, the proposed sparse grid integration method is applied to fit the MSVb model to foreign- exchange rate series of Euro (EUR)/Turkish Lira (TRL) and US Dollar (USD)/Turkish Lira (TRL) from March 1, 2001 to September 30, 2015. Special days and holidays, when the market was closed, were excluded. In the resulting data, there are 3669 trading days for each series. The returns are defined by the log-difference of each series. The return series were first checked for AR(1) effects and those effects were removed from the series. Returns of two foreign-exchange rates are shown in Figure 1 where the co-movement of the returns is observable.

The MSVb model was fitted to the data using the proposed sparse grid integration method where the trapezoid rule of level l = 8 is used as the basis univariate integration formula to construct the sparse grid as described in Section 3.1. The resulting two dimensional sparse grid has 1793 points for each time period with 255 points at each dimension. For identifying the integration intervals, extended Kalman filter is used and the integration interval for each time period t was set as described in Section 3.2, where v = 6.0 is used for the interval $\left(\hat{h}_{t|s} - v\hat{d}_{t|s}, \hat{h} + v\hat{d}_{t|s}\right)$. For the parameter estimation, log-likelihood given in Equation (18) was maximized using the Newton's method. The standard errors and confidence intervals on parameters are obtained from the Hessian produced by the Newton's methods as explained in Section 3.3.

Table 3 summarizes the parameter estimation results on the foreign exchange data. The MSVb model estimated by the sparse grid integration method successively captures the high persistence of the log-volatilities ($\varphi_1 = 0.971$, $\varphi_2 = 0.971$) and the correlations between returns ($\rho_{12} = 0.649$).

As it can be seen in Table 4, the MSVb model addresses the patterns of log-volatilities in the logarithms of squared returns better than the approximation with extended Kalman filter with lower RMSEs for all estimation types namely filtering, smoothing and one-step ahead predictions.

Figure 2 shows the smoothing estimates of log-volatilities obtained by the sparse grid integration and the Kalman filter methods, along with the logarithms of squared returns as benchmark, where better coherence of sparse grid integration estimation patterns is visually supported.

7. CONCLUSIONS

The results obtained from the simulated data and empirical application show that the estimation algorithms based on the proposed sparse grid integration method perform well enough to be considered as an alternative estimation method for the MSV models. Sequential algorithmic nature, highly parallelizable structure, better error control and convergence properties of sparse grid integration based estimation algorithms, compared to the batch nature of MCMC method, their limited parallelizability capabilities, and convergence issues, make sparse grid integration based estimation algorithms quite promising, especially for practical applications. Moreover, for the estimation problems of filtering and prediction, sparse grid integration based algorithms achieve better performance than MCMC based algorithms in terms of both accuracy and computational requirements. Filtering and prediction are as important as smoothing and parameter estimation, especially in practical applications, so the proposed algorithms based on the sparse grid integration provide the required versatility.

Further research includes construction of sparse grid formulas from other univariate formulas such as Gaussian quadrature rules which may and probably be more effective and suitable to MSV model density structures. Another direction for improvement would be the construction strategy of the sparse grid; a dynamic construction of the sparse grid by adjusting the level of integration formula at each time step for better error control and computational efficiency can significantly improve the method from the algorithmic perspective. Hybrid approaches combining Monte Carlo based methods with sparse grid integration based methods for computational advantage and better convergence could lead to further research.

Table 3. Estimated MSVb model parameters for EUR/TL and USD/TL returns.

Parameter	Series	Mean	Standard error	95% interval
γ_i	1	−0.289	0.027	[−0.342, −0.237]
	2	−0.378	0.033	[−0.440, −0.314]
φ_i	1	0.971	0.0092	[0.954, 0.987]
	2	0.962	0.0084	[0.946, 0.976]
$\sigma_{\eta i}$	1	0.188	0.019	[0.154, 0.225]
	2	0.184	0.023	[0.146, 0.229]
ρ_{ij}	1,2	0.649	0.039	[0.572, 0.725]

a. EUR/TL is Series 1 and USD/TL is Series 2.

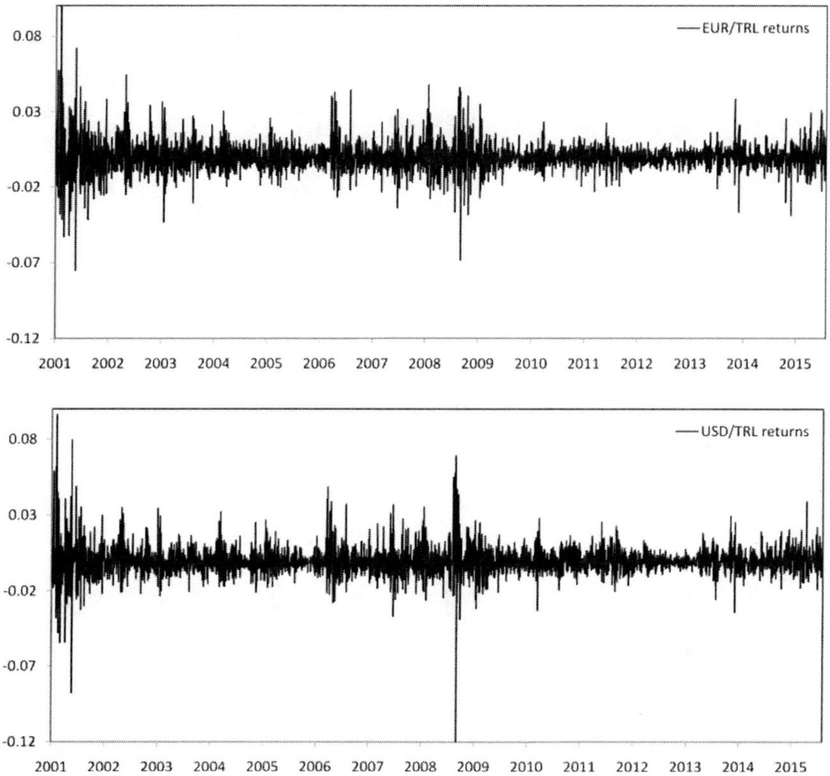

Figure 1. Returns of USD/TRL and EUR/TRL.

Table 4. Comparison of MSVb model estimates using sparse-grid integration and extended Kalman filter.

	i	RMSEs for SV with sparse grid integration			RMSEs for extended Kalman filter		
		Filtering, $h_{t\|t}$	Smoothing, $h_{t\|T}$	Prediction, $h_{t+1\|t}$	Filtering, $h_{t\|t}$	Smoothing, $h_{t\|T}$	Prediction, $h_{t+1\|t}$
h_i	1	2.576	2.436	2.660	2.711	2.652	2.754
	2	2.583	2.442	2.687	2.691	2.644	2.751

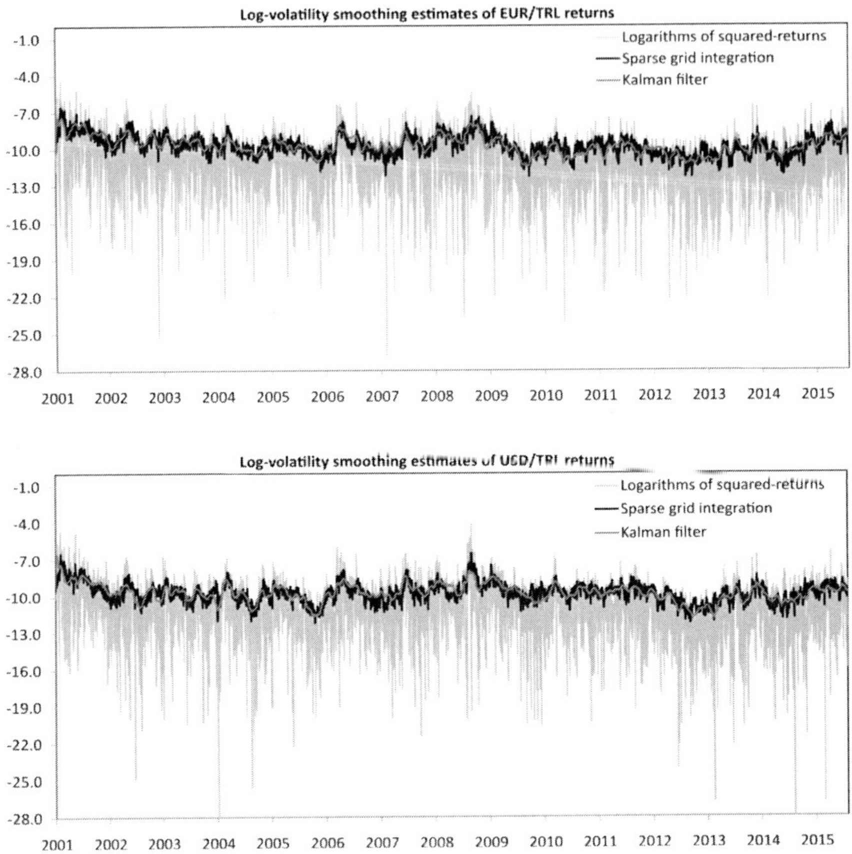

Figure 2. Smoothing estimates of log-volatilities.

REFERENCES

1. 1. Xiao, L. and Abdurrahman, A. (2007) Volatility Modelling and Forecasting in Finance. In: Knight, J. and Satchell, S., Eds., Volatility in the Financial Markets, Elsevier, 1-45.
2. Shephard, N. and Andersen, T.G. (2008) Stochastic Volatility: Origins and Overview. Discussion Paper Series, University of Oxford Department of Economics, Oxford.

3. Poon, S.H. and Granger, C.J.W. (2003) Forecasting Volatility in Financial Markets: A Review. Journal of Economic Literature, 41, 478-539.

4. Tsay, R.S. (2002) Analysis of Financial Time Series: Financial Econometrics. John Wiley, New York.

5. Ghysels, E., Harvey, A. and Renault, E. (1996) Stochastic Volatility. In: Maddala, G.S. and Rao, C.R., Eds., Handbook of Statistics (14) Statistical Methods in Finance, Elsevier, Amsterdam, 119-191.

6. Taylor, S.J. (1982) Financial Returns Modelled by the Product of Two Stochastic Processes—A Study of Daily Sugar Prices 1961-79. In: Anderson, O.D., Ed., Time Series Analysis: Theory and Practice 1, North-Holland, Amsterdam, 203-226.

7. Taylor, S.J. (1986) Modelling Financial Time Series. John Wiley, Chichester.

8. Harvey, A., Ruiz, E. and Shephard, N. (1994) Multivariate Stochastic Variance Models. Review of Economic Studies, 61, 247-264.

9. Asai, M. and McAleer, M. (2006) Asymmetric Multivariate Stochastic Volatility. Econometric Reviews, 25, 453-473.

10. Asai, M. and McAleer, M. (2009) The Structure of Dynamic Correlations in Multivariate Stochastic Volatility Models. Journal of Econometrics, 150, 182-192.

11. Ishihara, T. and Omori, Y. (2012) Efficient Bayesian Estimation of a Multivariate Stochastic Volatility Model with Cross-Leverage and Heavy-Tailed Errors. Computational Statistics and Data Analysis, 56, 3674-3689.

12. Jiang, G., Knight, J. and Rossi, P.E. (2005) Alternative Specifications of Stochastic Volatility Models. Working Paper, University of Western Ontario, Ontario.

13. Watanabe, T. (1999) A Non-Linear Filtering Approach to Stochastic Volatility Models with an Application to Daily Stock Returns. Journal of Applied Econometrics, 14, 101-121.

14. Tanizaki, H. (1997) Nonlinear and Nonnormal Filters Using Monte Carlo Methods. Computational Statistics & Data Analysis, 25, 417-439.

15. Harvey, A. and Shephard, N. (1996) Estimation of an Asymmetric Stochastic Volatility Model for Asset Returns. Journal of Business and Economic Statistics, 14, 429-434.

16. Galant, A.R., Hsieh, D.A. and Tauchen, G. (1997) Estimation of Stochastic Volatility Models with Diagnostics. Journal of Econometrics, 81, 159-192.

17. Carlin, B.P., Polson, N.G. and Stoffer, D.S. (1992) A Monte Carlo Approach to Nonnormal and Nonlinear State Space Modeling. Journal of American Statistical Association, 87, 493-500.

18. Sandman, G. and Koopman, S.J. (1998) Estimation of Stochastic Volatility Models via Monte Carlo Maximum Likelihood. Journal of Econometrics, 87, 271-301.

19. Tanner, M.A. and Wong, W.H. (1987) The Calculation of Posterior Distributions by Data Augmentation. Journal of the American Statistical Association, 82, 528-550.

20. Tierney, L. (1994) Markov Chains for Exploring Posterior Distributions. The Annals of Statistics, 22, 1701-1762.

21. Chib, S. and Greenberg, E. (1995) Understanding the Metropolis-Hastings Algorithm. The American Statistician, 49, 327-335.

22. Chib, S. and Greenberg, E. (1996) Markov Chain Monte Carlo Simulation Methods in Econometrics. Econometric Theory, 12, 409-431.

23. Jacquier, E., Polson, N.G. and Rossi, P.E. (1994) Bayesian Analysis of Stochastic Volatility Models. Journal of Business and Economic Statistics, 12, 371-417.

24. Kitagawa, G. (1987) Non-Gaussian State Space Modeling of Nonstationary Time Series. Journal of American Statistics Association, 82, 1032-1063.

25. Smolyak, S.A. (1963) Quadrature and Interpolation Formulas for Tensor Products of Certain Classes of Functions. Soviet Mathematics Doklady, 4, 240-243.

26. Heiss, F. and Winschel, V. (2006) Estimation with Numerical Integration on Sparse Grids. Technical Report, University of Munich, Munich.

27. Bungarts, H.J. and Griebel, M. (2004) Sparse Grids. Acta Numerica, 13, 147-270.

28. Gerstner, T. and Griebel, M. (1998) Numerical Integration Using Sparse Grids. Numerical Algorithms, 18, 209-232.

29. Gerstner, T. (2007) Sparse Grid Quadrature Methods for Computational Finance. University of Bonn, Habilitation.

30. Holtz, M. (2010) Sparse Grid Quadrature in High Dimensions with Applications in Finance and Insurance. Vol. 77, Springer Science & Business Media, Berlin.

31. Rebonato, R. and J&aumlckel, P. (2011) The Most General Methodology to Create a Valid Correlation Matrix for Risk Management and Option Pricing Purposes.

32. Kercheval, A.N. (2008) On Rebonato and Jackel's Parametrization Method for Finding Nearest Correlation Matrices. International Journal of Pure and Applied Mathematics, 45, 383-390.

33. Rapisarda, F., Brigo, D. and Mercurio, F. (2007) Parameterizing Correlations: A Geometric Interpretation. IMA Journal of Management Mathematics, 18, 55-73.

9
CHAPTER

Numerical Approximation of Fractal Dimension of Gaussian Stochastic Processes

Freddy H. Marin Sanchez, William Eduardo Alfonso

Basic Science Department, EAFIT University, Carrera 49 # 7 Sur-50, Medellín, Colombia

ABSTRACT

In this paper we propose a numerical method to estimate the fractal dimension of stationary Gaussian stochastic processes using the random Euler numerical scheme and based on an analytical formulation of the fractal dimension for filtered stochastic signals. The discretization of continuous time processes through this random scheme allows us to find, numerically, the expected value, variance and correlation functions at any point of time. This alternative method for estimating the fractal dimension is easy to implement and requires no sophisticated routines. We use simulated data sets for stationary processes of the type Random Ornstein Uhlenbeck to graphically illustrate the results and compare them with those obtained whit the box counting theorem.

Keywords: Stationary Gaussian Stochastic Processes, Fractal Dimension, Random Euler Numerical Scheme

1. INTRODUCTION

A common practice when trying to measure irregular shapes, such as the perimeter of an island or the length of a coastline, is to use Euclidean geometry, ignoring the fact that these shapes do not correspond to the ones of ideal objects, such as polygons and circles, whose dimension is an integer. In contrast, many forms observed in nature have a special feature that makes their resulting measurement depend on the measuring scale, such that the lower the scale, the higher the value of the measurement, to the point where it becomes infinite or indeterminate. The dimension of irregular shapes is a non-integer dimension known as the Hausdorff-Besicovitch dimension. In the early twentieth century, Lewis Fry Richardson found a relationship that allows determining the value of a constant that indicates the degree of roughness of a coast or a geographical border, and used it to calculate the border length between several countries [1] . Based on Richardson's findings, Benoit Mandelbrot [2] found that such irregular shapes have the property of statistical self-similarity, which states that any portion of an irregular shape can be considered a fully scaled image of the entire shape. Later, Mandelbrot coined this property as the fractal dimension to describe sets in $\mathbb{R}^d$. There are many studies in the literature in which maps, border areas and natural and physical phenomena are modeled and simulated with fractal features. In [3] for example, a proposal to simulate real maps using the fractal dimension of Gaussian stochastic processes is presented and some reports modeling terrestrial surfaces, irregular oceanic coasts and rivers, can be found in [4] -[9] .

There are several approaches to estimating of the fractal dimension. In fact in [10] different types of fractals that appear in scientific research are examined, application in nonlinear dynamical systems is discussed, and an exahustive review of numerical methods for their estimation is done.

[11] proposed another method for estimating the fractal dimension of subset of $\mathbb{R}^d$, as well as an additional correction to address the problem of resolution, allowing obtaining not only the fractal dimension, but also a confidence interval for this estimation, thus improving the box counting method. They also show some

experimental results based on simulated data and real data from the coastline of Norway. The techniques for the measurement of fractal dimension can be divided into two groups: those based on vectors, which include structured algorithms, and those based on matrices, for the analysis of image systems that include the distance transformation method reported in [12] . Moreover, [13] introduces the roughness method in order to determine the fractal dimension of the Sierpinski carpet, triangle and tetrahedron, the standard Cantor set and the Menger sponge. Meanwhile, in the study of stochastic signals filtering by [14] is obtained the analytical formulation of the fractal dimension to investigate the effects of a linear filter and the regularity in a particular mono-fractal stochastic process, based on an explicit, continuous formulation, linking the spectral properties of a given filter and the fractal properties of a given process, knowing the function of probability density and output, departing from the correlation coefficient of the filter. With regard to the arguments established by [14] , the aim of our study is to estimate, numerically, the fractal dimension of ergodic stationary Gaussian stochastic processes using the random Euler scheme.

This paper is organized as follows. Section 2 presents the Euler random scheme and builds the numerical solution, the expected value and the crossed expected value for a more general stochastic processe that allows calculating the variance and correlation of the solution of problems of random initial value; the definition of fractal dimension and the box counting theorem are described in Section 3. In Section 4 we present a numericalanalytical formulation for calculating length and fractal dimension for stationary Gaussian stochastic processes. A method to calculate the length and fractal dimension of stationary Gaussian stochastic processes that include processes such as random Ornstein Uhlenbeck is presented in Section 5. Section 6 illustrates graphically, some experimental numerical results associated with the calculation of the fractal dimension of ergodic stationary Gaussian stochastic processes, compared with the results obtained from using the box count theorem.

2. ON THE RANDOM EULER NUMERICAL SCHEME

Consider the initial random value problem of the form[1],

$$\dot{X}(t) = f(X(t),t), \quad t \in T = [t_0, t_1]$$
$$X(t_0) = X_0 \qquad (1)$$

where X_0 is a random variable and therefore $X(t)$ as $f(X(t),t)$ are both unknown stochastic processes defined on the same probability space $(\Omega, \mathcal{F}, \mathbb{P})$. The aim of this section is to use the Euler scheme to construct a numerical solution, the expected value and the crossed expected value to random initial value problem (1).

Note that the discretized solution of Equation (1) can be expressed in integral form as

$$X_{t_n} - X_{t_{n-1}} = \int_{t_{n-1}}^{t_n} f(X(s), S)\,ds.$$

By using the approximation $\int_{t_{n-1}}^{t_n} f(S, X(s))\,ds \approx f(t_{n-1}, X(t_{n-1}))(t_n - t_{n-1})$, the numerical scheme

$$\begin{cases} X_{t_n} = X_{t_{n-1}} + hf(X_{t_{n-1}}, t_{n-1}) & n \geq 1, \\ X_{t_0} = X_0 \end{cases} \qquad (2)$$

is obtained, where X_{t_n}, $f(X_{t_{n-1}}, t_{n-1})$ are second order random variables (2-r.v.) and $h = t_n - t_{n-1}$.

In particular, consider a (2-r.v.) X_0 and a stochastic process $b(t)$ second order (2-s.p.) m.s. integrable defined on the same probability space $(\Omega, \mathcal{F}, \mathbb{P})$ such that $b(t)$ is independent of X_0 for each t lying in the interval $T=[t_0,t_1]$. Let $a(t)$ a continuous deterministic function defined on T. The idea is to solve numerically initial random value problems given by the expression

$$\begin{aligned} \dot{X}(t) &= a(t)X(t)+b(t), \quad t\in T=[t_0,t_1] \\ X(t_0) &= X_0. \end{aligned} \tag{3}$$

Note that the Euler numerical scheme for problem (3)[2] takes the form

$$\begin{aligned} X_{t_n} &= X_{t_{n-1}} + h\left[a(t_{n-1})X_{t_{n-1}} + b(t_{n-1})\right], \\ &= \left(1+ha(t_{n-1})\right)X_{t_{n-1}} + hb(t_{n-1}) \quad n\geq 1. \end{aligned} \tag{4}$$

It is easy to show using the principle of mathematical induction, (see [19] and [20] pp. 58-59), that X_{t_n} and $X_{t_n}X_{t_m}$ can be expressed as

$$X_{t_n} = X_{t_0}\prod_{i=0}^{n-1} f_i + h\sum_{i=0}^{n-1}\left(\prod_{j=i+1}^{n-1} f_j\right)b(t_i); \qquad n\geq 1. \tag{5}$$

$$\begin{aligned} X_{t_n}X_{t_m} &= X_{t_0}^2\left(\prod_{i=0}^{n-1} f_i\right)\left(\prod_{j=0}^{m-1} f_j\right) + h^2\sum_{i=0}^{n-1}\sum_{k=0}^{m-1}\left(\prod_{j=i+1}^{n-1} f_j\right)\left(\prod_{l=k+1}^{m-1} f_l\right)b(t_i)b(t_k) \\ &+ hX_{t_0}\left[\left(\prod_{i=0}^{n-1} f_i\right)\sum_{k=0}^{m-1}\left(\prod_{l=k+1}^{m-1} f_l\right)b(t_k) + \left(\prod_{k=0}^{m-1} f_k\right)\sum_{i=0}^{n-1}\left(\prod_{j=k+1}^{n-1} f_j\right)b(t_i)\right], \end{aligned} \tag{6}$$

where $f_i = 1+ha(t_i)$. From Equations (5), (6) and the results reported in [19] , it follows that the expected values $E\{X_{t_n}\}$ and $E\{X_{t_n}X_{t_m}\}$ takes the form

$$E\left\{X_{t_n}\right\} = E\left\{X_{t_0}\right\}\left(\prod_{i=0}^{n-1} f_i\right) + h\sum_{i=0}^{n-1}\left(\prod_{j=i+1}^{n-1} f_j\right)E\left\{b(t_i)\right\}, \tag{7}$$

$$\begin{aligned} E\left\{X_{t_n}X_{t_m}\right\} &= E\left\{X_{t_0}^2\right\}\left(\prod_{i=0}^{n-1} f_i\right)\left(\prod_{i=0}^{n-1} f_k\right) + h^2\sum_{i=0}^{n-1}\sum_{k=0}^{m-1}\left(\prod_{j=i+1}^{n-1} f_j\right)\left(\prod_{l=k+1}^{m-1} f_l\right)E\left\{b(t_i)b(t_k)\right\} \\ &+ hE\left\{X_{t_0}\right\}\left[\left(\prod_{i=0}^{n-1} f_i\right)\sum_{k=0}^{m-1}\left(\prod_{l=k+1}^{m-1} f_l\right)E\left\{b(t_k)\right\} + \left(\prod_{k=0}^{m-1} f_k\right)\sum_{i=0}^{n-1}\left(\prod_{l=k+1}^{n-1} f_j\right)E\left\{b(t_i)\right\}\right]. \end{aligned} \tag{8}$$

3. FRACTAL DIMENSION AND THE BOX COUNTING THEOREM

The fractal dimension is an objective way to compare fractals geometrically. This section introduces the concept of fractal dimension and presents the box counting theorem.

Let (X,d) denote a complete metric space. Let $A \in H(X)$ be a nonempty compact subset of X. Let $B(x,\epsilon)$ denote the closed ball of radius $\epsilon > 0$ and center at a point $x \in X$. We wish to define an integer, $N(A,\epsilon)$, to be the least number of closed balls of radius ϵ needed to cover the set A.

That is, $N(A,\epsilon) = M$: smallest positive integer M such that $A \subset \bigcup_{n=1}^{M} B(x_n,\epsilon)$, for some set of distinct points $\{x_n : n = 1,2,\cdots,M\} \subset X$. In other words, surround every point $x \in A$ by an open ball of radius $\epsilon > 0$ to provide a cover of A by open sets. Because A is compact this cover possesses a finite subcover, consisting of an integer number, say $\hat{M}$, of open balls. By taking the closure of each ball, thus obtained a cover consisting of M closed balls. Let C denote the set of covers of A by at most $\hat{M}$ closed balls of radius ϵ. Then C contains at least one element.

Let $f:C\rightarrow\{1,2,3,\cdots,M\}$ defined by $f(c)=$ number of balls in the cover $c\in C$. Then $\{f(c):c\in C\}$ is a finite set of positive integers. It follows that it contains a least integer, $N(A,\epsilon)$.

The intuitive idea behind fractal dimension is that a set A has fractal dimension D if: $N(A,\epsilon)\approx C\epsilon^{-D}$ for some positive constant C 3.

Accordingly,

$$D\approx\frac{\ln N(A,\epsilon)-\ln C}{\ln(1/\epsilon)}.$$

Now notice that the term $\frac{\ln C}{\ln(1/\epsilon)}$ approaches 0 as $\epsilon\rightarrow 0$. This leads us to the following definition.

Definition 3.1. Let $A\in H(X)$ where (X,d) is a metric space. For each $\epsilon>0$, let $N(A,\epsilon)$ denote the smallest number of closed balls of radius $\epsilon>0$ needed to cover A. If

$$D=\lim_{\epsilon\rightarrow 0}\left\{\frac{\ln N(A,\epsilon)}{\ln(1/\epsilon)}\right\}$$

exists, then D is called the fractal dimension A. Also used the notation $D=D_A$ and will say " A has fractal dimension D ".

Theorem 3.2. The Box Counting Theorem. Let $A\in H(\mathbb{R}^m)$, where the Euclidean metric is used. Cover $\mathbb{R}^m$ by closed square boxes of side length $(1/2^n)$. Let $N_n(A)$ denote the number of boxes of side length $(1/2^n)$ which intersect the attractor. If

$$D = \lim_{n \to \infty} \left\{ \frac{\ln(N_n(A))}{\ln(2^n)} \right\},$$

then A has fractal dimension D. The proof of this theorem and other important definitions can be found at [21].

4. ANALYTICAL FORMULATION OF THE FRACTAL DIMENSION FOR STOCHASTIC PROCESSES

An analytical expression for the fractal dimension of filtered stochastic signals with the purpose of studying the effects of a linear filter and the regularity of a given stochastic mono-fractal process is introduced in [14] . They present some arguments using the Euclidean length of a deterministic curve $x(t)$ with limited duration T, as starting point to estimate the length and fractal dimension of such stochastic signals.

However, stochastic processes can not be described by deterministic continuous curves and irregularities are present even when examined with higher resolutions. This leads to the paradoxical conclusion that the lengths of such signals are not finite. To overcome the fact that stochastic processes lack of a finite longitude, [2] assumes that the longitude L_x of a fractal curve depends on the resolution and the length of the measurement instrument, λ. [1] proves that measuring $L_x(\lambda)$ as a function of λ allows to determine the fractal dimension i.e.:

$$L_x(\lambda) \approx K_x \lambda^{1-D_x} \tag{9}$$

where D_x is the fractal dimension of the curve and K_x is a constant.

Thus, the length L_x can be written as a function of λ given by

$$L_x(\lambda) \approx T\overline{\lambda(h)} \tag{10}$$

where $\overline{\lambda(h)} = \frac{1}{T}\int_{-T/2}^{T/2} \lambda(t,h)\,\mathrm{d}t$ and where the elementary length is given by $\lambda(t,h) \approx (1/h)\sqrt{h^2 + \Delta(t,h)^2}$.

This definition of length for continuous curves can be extended to ergodic stationary stochastic processes by replacing the arithmetic mean by the mathematical expectation $E\{.\}$, i.e.

$$L_x(\lambda) = TE[\lambda]$$

where $\lambda = (1/h)\sqrt{h^2 + \Delta^2}$.

Note that $L_x(h) = \frac{T}{h}E\left\{\sqrt{h^2 + \Delta^2}\right\}$ only is solved numerically. In particular, for a high frequency sampling it can be assumed that $h \ll \Delta$ to obtain an analytical expression given by [22]

$$L_x(h) \approx \frac{T}{h}E\{|\Delta|\}. \tag{11}$$

Length and Fractal Dimension for Gaussian Stochastic Processes

Following the results reported in [14] , we will make a brief extension that also requires a prior information such as the probability density function of the process and the variance and correlation functions.

Consider a stochastic process $\{X_t\}_{t\geq 0}$ with values in $\mathbb{R}$ defined on a probability space $(\Omega, \mathcal{F}, \mathbb{P})$. The average length of the process can be estimated by the expression

$$L_x(t) \approx \frac{T}{h} E\{|x_1(t) - x_2(t)|\} \tag{12}$$

with $x_1(t) = X_t, x_2(t) = X_{t-h}; t, t-h \in [0,T]$. Thus, the expected value of the absolute value of the difference between $x_1(t)$ and $x_2(t)$ takes the form

$$E\{|x_1(t) - x_2(t)|\} = \iint |x_1(t) - x_2(t)|\, p_{x_1,x_2}(x_1(t), x_2(t))\, \mathrm{d}x_1 \mathrm{d}x_2, \tag{13}$$

where, $p_{x_1,x_2}(x_1(t), x_2(t))$ denotes the joint probability density function.

Now suppose that the stochastic process $\{X_t\}_{t\geq 0}$ is a stationary Gaussian process. Consequently, $x_1(t)$ and $x_2(t)$ are also Gaussian stationary and therefore the joint probability density can be written as

$$p_{x_1,x_2}(x_1(t), x_2(t)) = \frac{\left(1-\rho_x^2(t)\right)^{1/2}}{2\pi\sigma_1(t)\sigma_2(t)} \mathrm{e}^{\left(-1/\left(2\left(1-\rho_x^2(t)\right)\right)\right)\left(x_1^2(t)/\sigma_1^2(t) + x_2^2(t)/\sigma_2^2(t) - 2\rho_x(t)x_1(t)x_2(t)/\sigma_1(t)\sigma_2(t)\right)}, \tag{14}$$

where $\sigma_1^2(t)$ and $\sigma_2^2(t)$ are the variances of $x_1(t)$ and $x_2(t)$ respectively, and $\rho_x(t)$ is the correlation function defined by

$$\rho_x(t) = E\{x_1(t)x_2(t)\} \Big/ \sqrt{E\{x_1^2(t)\}E\{x_2^2(t)\}}.$$

The probability density of the increment $\Delta = x_1(t) - x_2(t)$ is

$$p_\Delta(\Delta) = \int_{-\infty}^{+\infty} p_{x_1x_2}\left(x_1(t), x_1(t)+\Delta\right)\mathrm{d}x_1. \tag{15}$$

Analogously to [14] , the probability density of the increment follows a normal distribution $N\left(0, \tilde{\sigma}(t)^2\right)$:

$$p_\Delta(\Delta) = \frac{1}{\sqrt{2\pi\tilde{\sigma}(t)}} \mathrm{e}^{-\Delta^2/2\tilde{\sigma}(t)^2}, \tag{16}$$

with $\tilde{\sigma}(t) = \sqrt{\sigma_1^2(t) + \sigma_2^2(t) - 2\rho_x(t)\sigma_1(t)\sigma_2(t)}$. As $x_1(t)$ and $x_2(t)$ belong to the same process $\{X_t\}_{t\geq 0}$ then $\sigma_1(t) = \sigma_2(t) = \sigma_x(t)$ and the standard deviation is $\tilde{\sigma}(t) = \sigma_x(t)\sqrt{2\left(1-\rho_x(t)\right)}$.

The mathematical expectation of the increment becomes

$$E\{|\Delta|\} = 2\int_0^{+\infty} \Delta p_\Delta(\Delta)\mathrm{d}\Delta = \sqrt{\frac{2}{\pi}}\tilde{\sigma}(t), \tag{17}$$

and thus

$$L_x(t) \approx \frac{2}{\sqrt{\pi}} \frac{T\sigma_x(t)}{h} \sqrt{1-\rho_x(t)}; \quad \forall t \in [0, T]. \tag{18}$$

Using Equation (9), the fractal dimension D_x of the process $\{X_t\}_{t\geq 0}$ can be approximated as

$$D_x(t,h) \approx 1 - \frac{\log\left(L_x(t)\right)}{\log(h)} + \frac{\log(K)}{\log(h)}. \tag{19}$$

Finally, Equation (18) becomes

$$D_x(t,h)=2-\frac{1}{2}\lim_{h\to 0}\frac{\log(1-\rho_x(t))}{\log(h)}. \tag{20}$$

5. NUMERICAL APPROACH TO FRACTAL DIMENSION FOR STATIONARY GAUSSIAN PROCESSES

In this section we present the numerical construction of the fractal dimension for stationary Gaussian stochastic processes including random Ornstein Uhlenbeck process, using the results of Section 4, the random Euler numerical scheme and the approximation of correlation function using the calculations for $E\{X_{t_n}\}$ and $E\{X_{t_n}X_{t_m}\}$ described in Section 2.

5.1. Calculation of Fractal Dimension for the Random Ornstein Uhlenbeck Process

Consider the random initial value problem given by

$$\dot{X}(t)=-\alpha X(t)+\sigma W(t);\ t\in[0,T] \tag{21}$$

$$X(0)=X_0$$

where $\alpha>0$, $\sigma>0$, and $W(t)$ is a Gaussian white noise independent of X_0 for each t lying in the interval $[0,T]$ (defined as in [16] p. 121) and $X_0\sim N\left(0,\frac{\sigma^2}{2\alpha}\right)$. Note that this process is stationary Gaussian and ergodic, and can be considered as a modification of the random walk in continuous time and analog to the order 1 autoregressive process,

AR (1) with discrete time, that broadly describes the velocity of a Brownian mass particle under the influence of friction [23] .

Under random Euler numerical scheme (5), the Equation (21) can be expressed as

$$X_{t_n} = \prod_{i=0}^{n-1}(1-\alpha h)X_0 + \sigma h\sum_{i=0}^{n-1}\left(\prod_{j=i+1}^{n-1}(1-\alpha h)\right)W(t_i); \quad 1 \le n \le N; h = \frac{T}{N}. \tag{22}$$

$$X_{t_n} = \beta^n X_0 + \sigma\sqrt{h}\sum_{i=0}^{n-1}\beta^{n-1-i}\xi_i, n \ge 1, \quad \xi_i \sim N(0,1),$$

where $\beta = 1-\alpha h$. By Equation (8) it follows that

$$E\{X_{t_n}X_{t_m}\} = \beta^{n+m}E\{X_0^2\} + \sigma^2 hE\left\{\sum_{i=0}^{n-1}\beta^{n-1-i}\xi_i\sum_{j=0}^{m-1}\beta^{m-1-j}\xi_j\right\}.$$

Trivially note that $E\{X_{t_n}\} = 0$. For $\beta < 1$, $\sum_{i=0}^{n-1}\beta^i = \frac{1-\beta^n}{1-\beta}$; thus

$$E\{X_{t_n}X_{t_m}\} = \beta^{n+m} + \sigma^2 h\left(\frac{1-\beta^{2\min\{n,m\}}}{1-\beta^2}\right).$$

For the case in which $n = N$ and $m = N-1$ it follows that

$$E\{X_{t_N}X_{t_{N-1}}\} = \beta^{2N-1} + \sigma^2 h\left(\frac{1-\beta^{2(N-1)}}{1-\beta^2}\right) = \beta E\{X_{t_{N-1}}^2\}.$$

It is easy to see that $\lim_{N\to\infty} E\{X_{t_N}^2\} = \lim_{N\to\infty} E\{X_{t_{N-1}}^2\} = \frac{\sigma^2 h}{1-\beta^2}$,

and $\lim E\{X_{t_N}X_{t_{N-1}}\} = \beta\frac{\sigma^2 h}{1-\beta^2}$. In this way,

$$\rho(h)=\frac{\beta\frac{\sigma^2 h}{1-\beta^2}}{\sqrt{\left(\frac{\sigma^2 h}{1-\beta^2}\right)\left(\frac{\sigma^2 h}{1-\beta^2}\right)}}=\beta,$$

which by Equation (20) becomes

$$D_x(h)=2-\frac{1}{2}\lim_{h\to 0}\frac{\log(\alpha h)}{\log(h)}. \tag{23}$$

5.2. Calculation of Fractal Dimension for a Random Mean Reverting Process

Consider the random initial value problem given by

$$\dot{X}(t)=\alpha(\mu-X(t))+\sigma W(t);\ t\in[0,T] \tag{24}$$

$$X(0)=X_0,$$

where $\alpha>0$, $\sigma>0$, $\mu\in\mathbb{R}$, $W(t)$ is a Gaussian white noise independent of X_0 for each t lying in the interval $[0,T]$ (defined as in [16] p. 121) and $X_0\sim N\left(0,\frac{\sigma^2}{2\alpha}\right)$ with $\sigma^2>2\alpha\mu^2$. Note that this process is stationary Gaussian and ergodic. When the initial condition is constant, this process is also known as mean reversion process with additive noise. This kind of process has been widely used in modeling the dynamic behavior of commodity prices and short term interest rates (see [24] and references therein), and the simulating of realistic irregular lattices in [3] .

Equation (24) can be written as

$$\dot{X}(t) = -\alpha X(t) + b(t); \quad b(t) = \alpha\mu + \sigma W(t). \tag{25}$$

Under random Euler numerical scheme, Equation (25) can be expressed as

$$X_{t_n} = \prod_{i=0}^{n-1}(1-\alpha h)X_0 + h\sum_{i=0}^{n-1}\left(\prod_{j=i+1}^{n-1}(1-\alpha h)\right)(\alpha\mu + \sigma W(t_i)); \quad 1 \le n \le N; h = \frac{T}{N}. \tag{26}$$

$$X_{t_n} = \theta\sum_{i=0}^{n-1}\beta^i + \beta^n X_0 + \sigma\sqrt{h}\sum_{i=0}^{n-1}\beta^{n-1-i}\xi_i, n \ge 1, \quad \xi_i \sim N(0,1).$$

Equation (8) implies that for $\theta = \alpha\mu h$ and $\beta = 1-\alpha h$,

$$E\{X_{t_n}X_{t_m}\} = \theta^2\sum_{i=0}^{n-1}\beta^i\sum_{j=0}^{m-1}\beta^j + \beta^{n+m}E\{X_0^2\} + \theta E\{X_0\}\left(\beta^m\sum_{i=0}^{n-1}\beta^i + \beta^n\sum_{j=0}^{m-1}\beta^j\right) + \sigma^2 hE\left\{\sum_{i=0}^{n-1}\beta^{n-1-i}\xi_i\sum_{j=0}^{m-1}\beta^{m-1-j}\xi_j\right\}.$$

Note that for $\beta < 1$, $\sum_{i=0}^{n-1}\beta^i = \frac{1-\beta^n}{1-\beta}$; in this way, $E\{X_n\} = \theta\left(\frac{1-\beta^n}{1-\beta}\right) + \beta^n E\{X_0\}$.

$$E\{X_{t_n}X_{t_m}\} = \theta^2\left(\frac{1-\beta^n}{1-\beta}\right)\left(\frac{1-\beta^m}{1-\beta}\right) + \beta^{n+m}E\{X_0^2\} + \theta E\{X_0\}\left(\beta^m\left(\frac{1-\beta^n}{1-\beta}\right) + \beta^n\left(\frac{1-\beta^m}{1-\beta}\right)\right) + \sigma^2 h\left(\frac{1-\beta^{2\min\{n,m\}}}{1-\beta^2}\right)$$

For the case in which $n = N$ and $m = N-1$ it follows that $E\{X_{t_N}X_{t_{N-1}}\} = \beta E\{X_{t_{N-1}}^2\} + \theta E\{X_{t_{N-1}}\}$.

It is easy to see that $\lim_{N\to\infty} E\{X_{t_N}\} = \frac{\theta}{1-\beta}$; $\lim_{N\to\infty} E\{X_{t_N}^2\} = \lim_{N\to\infty} E\{X_{t_{N-1}}^2\} = \frac{\theta^2}{(1-\beta)^2} + \frac{\sigma^2 h}{1-\beta^2}$ and

$$\lim_{N\to\infty} E\{X_{t_N}X_{t_{N-1}}\} = \beta\frac{\theta^2}{(1-\beta)^2} + \beta\frac{\sigma^2 h}{1-\beta^2} + \frac{\theta^2}{1-\beta}.$$

In this way,

$$\rho(h)=\frac{\beta\frac{\theta^2}{(1-\beta)^2}+\beta\frac{\sigma^2 h}{1-\beta^2}+\frac{\theta^2}{1-\beta}}{\sqrt{\left(\frac{\theta^2}{(1-\beta)^2}+\frac{\sigma^2 h}{1-\beta^2}\right)\left(\frac{\theta^2}{(1-\beta)^2}+\frac{\sigma^2 h}{1-\beta^2}\right)}}=\frac{\left(\mu^2+\frac{\sigma^2}{2\alpha-\alpha^2 h}\right)(1-\alpha h)+\alpha\mu^2 h}{\left(\mu^2+\frac{\sigma^2}{2\alpha-\alpha^2 h}\right)}=1-\frac{\alpha\sigma^2 h}{\sigma^2+\mu^2\left(2\alpha-\alpha^2 h\right)}$$

Equation (20) becomes

$$D_x(h)=2-\frac{1}{2}\lim_{h\to 0}\left[\frac{\log\left(\frac{\alpha\sigma^2 h}{\sigma^2+\mu^2\left(2\alpha-\alpha^2 h\right)}\right)}{\log(h)}\right]. \tag{27}$$

6. NUMERICAL RESULTS

This section graphically illustrates some experimental numerical results related to the calculation of the fractal dimension for stationary Gaussian stochastic processes associated with random initial value problems (21) and (24).

Figure 1(a) shows a comparison between the approximate average fractal dimension of 1000 simulated trajectories for Ornstein Uhlenbeck process using the Box Counting theorem and the numerical fractal dimension proposed in Equation (23) for different values of h. In Figure 1(b) the dynamic behavior of the fractal dimension is shown for different values of the parameter α .

Figure 2(a) shows a comparison between the approximate average fractal dimension of 1000 simulated trajectories for mean reverting process using the Box Counting theorem and the numerical fractal dimension proposed in Equation (27) for different values of h. In Figure 2(b) the dynamic behavior of the fractal dimension is shown for different values of the parameter α .

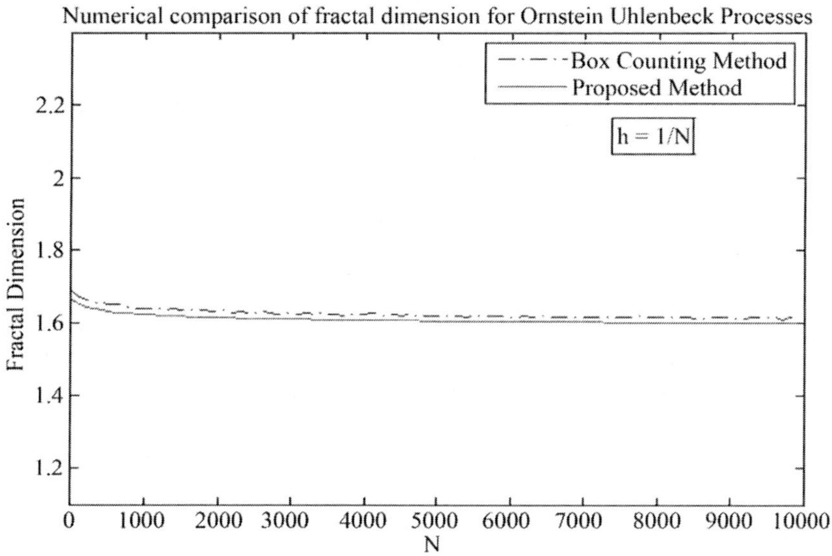

(a)

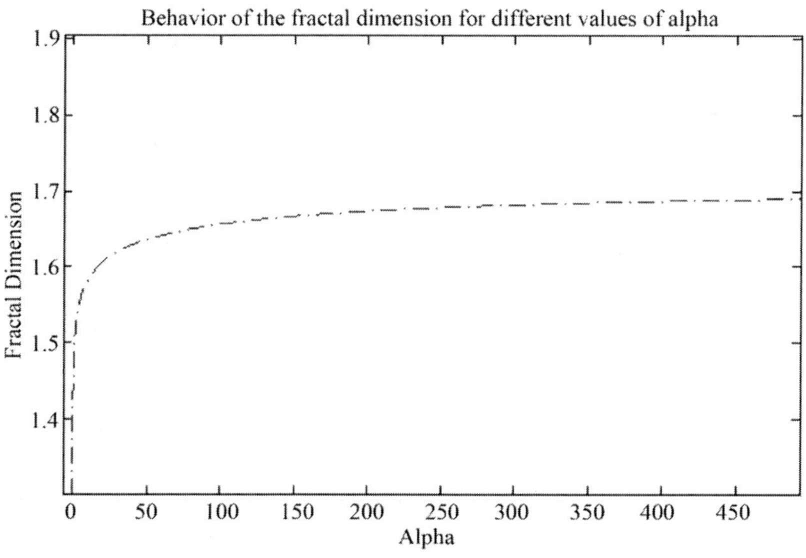

(b)

Figure 1. (a) Numerical comparison of fractal dimension for Ornstein Uhlenbeck processes with $T=1$, $\alpha=10$ and $\sigma=20$ for different values of h; (b) Dynamic behavior of the fractal dimension for Ornstein Uhlenbeck processes with $T=1$, $h=1/10^4$ and $\sigma=20$ for $0.001 \le \alpha \le 500$.

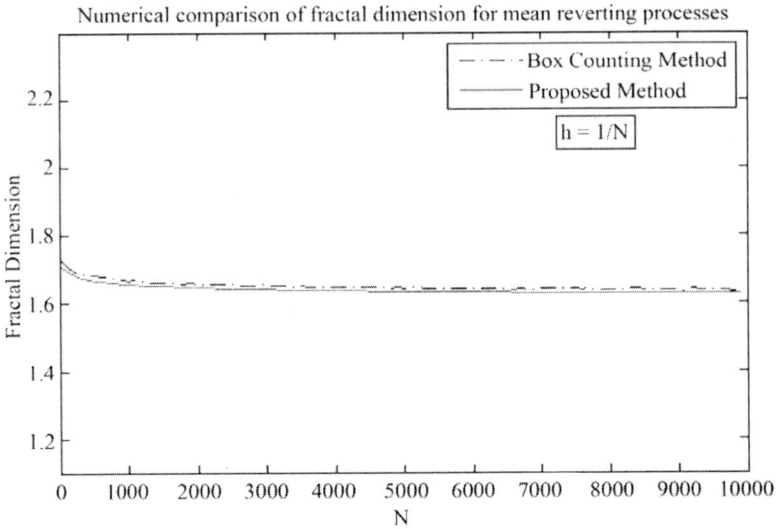

(a)

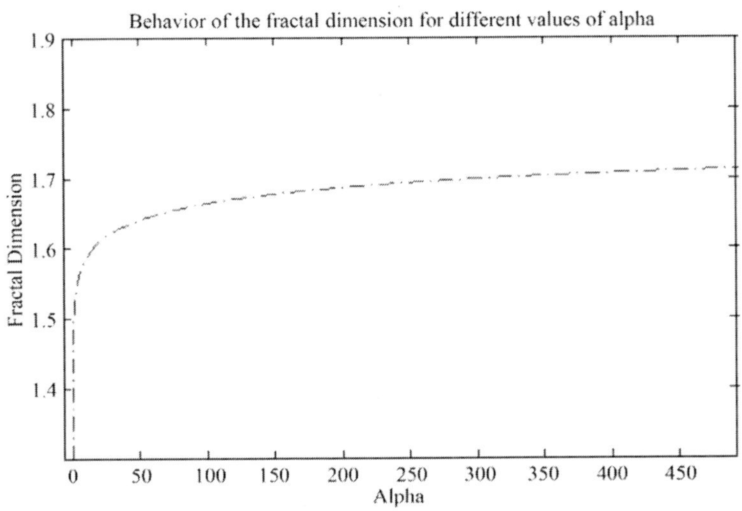

(b)

Figure 2. (a) Numerical comparison of fractal dimension for mean reverting processes with $T = 1$, $\mu = 10$, $\alpha = 20$ and $\sigma = 250$ for different values of h; (b) Dynamic behavior of the fractal dimension for mean reverting processes with $T = 1$, $\mu = 10$, $h = 1/10^4$ and $\sigma = 500$ for $0.001 \leq \alpha \leq 500$

.

7. CONCLUSIONS

This paper made a slight extension of the approach proposed in [14] for the numerical estimation of the fractal dimension of ergodic stationary Gaussian stochastic processes associated with the random Ornstein Uhlenbeck process. Using random Euler scheme performed numerical construction of the expectation value, variance and correlation functions at any point of time to initial random value problems general. The computational implementation of this alternative method to estimate the fractal dimension is easy and requires no sophisticated routines. It illustrates graphically the results obtained compared with box counting theorem using simulated data sets for stationary processes type random Ornstein Uhlenbeck.

A clear convergence of the proposed methods is shown in Figure 1(a) and Figure 2(a). In addition, the results show that the fractal dimension of Ornstein Uhlenbeck processes is independent of the parameter μ. This independence is maintained in the mean reverting process for the case where $\sigma^2 > 2\alpha\mu^2$. Figure 1(b) and Figure 2(b) reflect the fact that the fractal dimension of this type of process is more sensitive to the parameter α.

NOTES

[1]See for details of second-order random variables (2-r.v.), second-order stochastic processes (2-s.p.), and random differential equations.

[2]Some convergence criteria random Euler scheme in the mean square sense can be viewed at -.

[3]The notation $f(\epsilon) \approx g(\epsilon)$ as follows. $\lim_{\epsilon \to 0}\{\ln(f(\epsilon))/\ln(g(\epsilon))\} = 1$, where $f(\epsilon)$ and $G(\epsilon)$ be real valued functions of the positive real variable ϵ.

REFERENCES

1. Richardson, L. (1961) The Problem of Contiguity: An Appendix of Statistic of Deadly Quarrels. General Systems Year Book, 61, 139-187.

2. Mandelbrot, B. (1967) How Long Is the Coast of Britain Statistical Self-Similarity and Fractional Dimension. Science, 156, 636-638.

3. Duque, J.C., Betancourt, A. and Marin, F. (2014) An Algorithmic Approach for Simulating Realistic Irregular Lattices. In: Thill, J.C. and Dragicevic, S., Eds., GeoComputational Analysis and Modeling of Regional Systems, Springer Heidelberg, in press.

4. Mandelbrot, B. (1975) Stochastic Models for the Earth's Relief, the Shape and the Fractal Dimension of the Coastlines, and the Number-Area Rule for Islands. Proceedings of the National Academy of Sciences, 72, 3825-3828.

5. Adler, R.J. (1978) Some Erratic Patterns Generated by the Planar Wiener Process. Advances in Applied Probability, 10, 22-27.

6. Cullin, W.E. and Datko, M. (1987) The Fractal Geometry of the Soil-Covered Landscape. Earth Surface Processes and Landforms, 12, 369-385.

7. Cullin, W.E. (1988) Dimension and Entropy in the Soil-Covered Landscape. Earth Surface Processes and Landforms, 13, 619-648.

8. Beauvais, A. Montgomery, D. (1996) Influence of Valley Type on the Scaling Properties of River Planforms. Water Resources Research, 32, 1441-1448.

9. Montgomery, K. (1996) Sinuosity and Fractal Dimension of Meandering Rivers. Area, 28, 491-500.

10. Theiler, J. (1990) Estimating Fractal Dimension. JOSA A, 7, 1055-1073.

11. Taylor C. and Taylor, S.J. (1991) Estimating the Dimension of a Fractal. Journal of the Royal Statistical Society. Series B (Methodological), 53, 353-364.

12. Allen, M., Brown, G.J. and Miles, N.J. (1995) Measurement of Boundary Fractal Dimensions: Review of Current Techniques. Powder Technology, 84, 1-14.

13. Blachowski, A. and Ruebenbauer, K. (2009) Roughness Method to Estimate Fractal Dimension. Acta Physica Polonica A, 115, 636-640.

14. Girault, J., Kouame, D. and Ouahabi, A. (2010) Analytical Formulation of the Fractal Dimension of Filtered Stochastic Signals. Signal Processing, 90, 2690-2697.

15. Loeve, M. (1963) Probability Theory. D. Van N. Company, London.

16. Soong, T. (1973) Random Differential Equations in Science and Engineering. Academic Press, New York.

17. Jodar, L., Cortez, J. and Villafuerte, L. (2007) Mean Square Numerical Solution of Random Differential Equations: Facts and Possibilities. Computers and Mathematics with Applications, 53, 1098-1106.

18. Jodar, L. and Villafuerte, L. (2007) Numerical Solution of Random Differential Equations: A Mean Square Approach. Mathematical and Computer Modelling, 45, 757-765.

19. Marin, F. and Laniado, H. (2014) Convergence of Analytical Stochastic Processes in Mean Square. 1-9.

20. Agarwall, R. (1992) Difference Equations and Inequalities. Marcel Dekker, New York.

21. Barnsley, M. (1988) Fractals Everywhere. Academic Press, London.

22. Le Tavernier, E. (1998) La methode de Higuchi pour la dimension fractal. Signal Processing, 65, 115-128.

23. Uhlenbeck, G. and Ornstein, L. (1930) On the Theory of the Brownian Motion. Physical Review, 36, 823-841.

24. Marin, F. and Palacio, J. (2013) Gaussian Estimation of One-Factor Mean Reversion Processes. Journal of Probability and Statistics, 2013, 10 p.

10

CHAPTER

Neumann Series on the Recursive Moments of Copula-Dependent Aggregate Discounted Claims

Siti Norafidah Mohd Ramli * and Jiwook Jang

Department of Applied Finance and Actuarial Studies, Faculty of Business and Economics, Macquarie University, Sydney 2109, Australia

ABSTRACT

We study the recursive moments of aggregate discounted claims, where the dependence between the inter-claim time and the subsequent claim size is considered. Using the general expression for the m−thm-th order moment proposed by Léveillé and Garrido (*Scand. Actuar. J.* **2001**, *2*, 98–110), which takes the form of the Volterra integral equation (VIE), we used the method of successive approximation to derive the Neumann series of the recursive moments. We then compute the first two moments of aggregate discounted claims,*i.e.*, its mean and variance, based on the Neumann series expression, where the dependence structure is captured by a Farlie–Gumbel–Morgenstern (FGM) copula, a Gaussian copula and a Gumbel copula with exponential marginal distributions. Insurance premium calculations with their figures are also illustrated.

Keywords: aggregate discounted claims; moments; copulas; Volterra integral equation; Neumann series; insurance premium

1. INTRODUCTION

As the occurrence of catastrophe events becomes more frequent, the assumption of independence between event occurrence and claim severity is no longer sufficient in insurance risk modeling, given its impact on pricing and reserving, capital allocation solvency, as well as regulatory systems. The February 2009, Victorian bushfire in Australia (10,200 insurance claims amounting to approximately AUD 1.2 billion), the February 2011, Christchurch earthquake (USD 13 billion insured economic losses), the 2011 Great Eastern Japanese earthquake (loss amounting to as much as USD 40 billion), as well as the 2012 Hurricane Sandy (an expected loss of USD 25 billion) are the examples of this effect (see [1,2]).

In dealing with the dependency between the inter-claim arrivals and claim sizes, various approaches have been proposed in previous studies that can be noticed in [3,4,5,6,7,8,9,10,11], as well as the references therein. Regardless of the model used, we notice that previous research focused on either examining the expression of the moments of the aggregate discounted claims, Z(t)Z(t), as can be seen in [6,11,12,13,14], or by finding the related ruin measures and the ruin probability expressions, just like in [3,4,5,10,15].

Assuming the Poisson claim arrival process with claim sizes following mixed exponential distributions, [7] obtained the explicit expressions of the actuarial net premiums and the variances of the discounted aggregate claims from the Laplace transform of the distribution of the shot noise process, which was derived using the martingale approach. The first two moments of the aggregate discounted claims were obtained in [9] assuming the dependency between the claim sizes and the rates of claim occurrence affected by a Markovian environment, called the circumstance process. A delayed renewal process was also explored in [12,13,14,15], as well as [11] to accommodate the epochs between claim arrival and the observation of the risk process.

The asymptotical behaviour of a conditional tail probability dependence structure of claim sizes given the inter-claim arrival time was studied in [5,8]. Assuming that the conditional tail of claim size given the inter-claim time satisfies a certain condition for a bounded inter-claim time and a really huge claim size, [5] obtained the asymptotic tail probabilities of the discounted aggregate claims. Three copulas were indicated as fulfilling this assumption, which are the Farlie–Gumbel–Morgenstern (FGM) and the Frank and Ali–Mikhail–Haq (AMH) copulas, and the Weibull claim size was paired with exponential inter-claim arrival time in their numerical example. On the other hand, [8] explored the analytical properties related to the same dependence structure described by the survival copulas, such as their local and global uniformity.

Conditioning on the first arrival and using a renewal theory argument, [12] derived a useful expression for the m–thm-th recursive moment, whereby the inter-claim arrival time and the claim severity are assumed to be independent. The same conditioning argument was then applied in [6,10], assuming the FGM copula and then solved using the Laplace transform approach. More recently, [11] also adopted the same technique to derive the recursive moments of a Sparre Andersen risk process assuming a fairly general dependence structure between the inter-claim time and subsequent claim size variables, providing a simplified moments expression for assuming Erlang weights. Four types of copula were showcased in their examples of joint distribution between the said variables, which are the polynomial copula, the Bernstein copula, the generalized FGM copula, the extended FGM copula (references for these copulas are available in Section 3 of [11]).

The recursive moment equation resulting from the technique used in [6,12] takes the form of a Volterra integral equation of the second kind, which is widely used in the fields of mathematical physics, such as the electromagnetic and viscoelasticity fields, to represent the dynamics of materials that contain memory (refer to [16,17,18,19,20]). We are interested in using the same technique and then extend the recursive moments obtained in [6], so that it can be applied to any continuous bivariate distribution to accommodate the dependency between the two variables. To do so, we solve the recursive expression of the moments using the Neumann series obtained via the Picard method of successive approximations, upon which a selection of bivariate distributions can be applied, including bivariate copula.

This article is structured as follows. Section 2 will introduce the general framework of the continuous time renewal risk model together with its recursive moments with exponentially distributed inter-claim time and general claim size distribution. The dependency between the claim size and inter-claim time are then specified using a bivariate copula. For that purpose, we consider three copulas, which are the FGM copula, the Gaussian copula, which is a type of elliptical copula, and the Gumbel copula, an Archimedean type of copula, which is a natural candidate to represent an extreme value copula that caters for the one-sided dependence structure (see [21]).

In Section 3, we introduce the Volterra integral equation, which will be solved using the successive approximations method, leading to the Neumann series expression of the recursive moments, which is the main result of this paper. The Neumann series expression of the recursive moments allows the flexibility to capture various dependence structures provided by copula probability density functions (pdf).

Section 4 starts with the comparison between the value of moments obtained by our Neumann series expression assuming the FGM copula and the closed form solution by [6]. We then present the numerical analysis, showing the value of moments across the dependence parameter for each copula considered, assuming an exponentially distributed claim size. The illustration and comparison of moments, as well as premium values under the standard deviation principle are also included in this section. Section 5 concludes the article.

2. MODEL SETUP

We consider a continuous time renewal risk model as in [6], whereby $Z = \{Z(t)\}_{t\geq 0}$ with:

$$Z(t) = \begin{cases} \sum_{i=1}^{N(t)} e^{-\delta T_i} X_i & \text{if } N(t) > 0 \\ 0 & \text{if } N(t) = 0 \end{cases}$$

In this model, $N = \{N(t)\}_{t \geq 0}$ is a homogeneous Poisson process and X_i is a non-negative random variable (r.v.) representing the claim amount occurring at time T_i for $i = 1, 2, \ldots, N(t)$. The instantaneous rate of net interest, δ, is assumed to be deterministic.

We also define the inter-claim time variable r.v. W_j as:

$$W_j = \begin{cases} T_j & \text{for } j = 1 \\ T_j - T_{j-1} & \text{for } j = 2, 3, \ldots \end{cases}$$

The variables, X_j and W_j, are assumed to be continuous. In this study, we relax the independent assumption between the inter-claim time, WjWj, and the claim size, X_j, and we let $\{(X_j, W_j)\}_{j \in \mathbb{N}}$ to form a sequence of independent and identically distributed (i.i.d) random vectors, whose components are dependent.

2.1. Recursive Moments of Aggregate Discounted Claims

Conditioning on the arrival of the first claim as in [6,10,12], and knowing

that $\mathbf{E}(X^m | W = s) = \int_0^\infty x^m f_{X|W=s}(x) dx$ for $m \geq 1$, we have the general form of the m–th moments of aggregate discounted claim as the following:

$$\begin{aligned}
\mu_Z^{(m)}(T) = \mathbf{E}[Z^m(T)] &= \int_0^T f_W(s)e^{-m\delta s}\mathbf{E}(X^m|W=s)ds + \int_0^T f_W(s)e^{-m\delta s}\mu_Z^{(m)}(T-s)ds \\
&+ \sum_{j=1}^{m-1}\binom{m}{j}\int_0^T f_W(s)e^{-m\delta s}\mathbf{E}(X^j|W=s)\mu_Z^{(m-j)}(T-s)ds \\
&= \int_0^T\int_0^\infty e^{-m\delta s}x^m f_{X,W}(x,s)\,dxds + \int_0^T e^{-m\delta s}f_W(s)\mu_Z^{(m)}(T-s)ds \\
&+ \sum_{1\le j<m}\binom{m}{j}\int_0^T\int_0^\infty e^{-m\delta s}x^j f_{X,W}(x,s)\,\mu_Z^{(m-j)}(T-s)dxds
\end{aligned} \quad (1)$$

where $fX,W(x,s)$ is the bivariate probability density function (pdf) of the pair, X_j and W_j.

In this study, the joint pdf is described via a copula, $C_\theta(u,v)$, whose pdf is given by $c_\theta(u,v) = \frac{d^2}{dudv}C_\theta(u,v)$ with dependence parameter θ (see, e.g., [22,23] for a general review on copulas). The bivariate pdf of (X,W) at (x,s) can be represented as:

$$f_{X,W}(x,s) = c_\theta(F_X(x),F_W(s))f_X(x)f_W(s)$$

where f(·) and F(·) are the marginal pdf and cdf for r.v.'s X and W.

Since the jump occurrences are assumed to follow a Poisson distribution, we therefore have an exponentially distributed inter-claim arrival time, $i.e., W \sim exp(\beta)$. Upon replacing $f_W(s) = \beta e^{-\beta s}$, we obtain:

$$\begin{aligned}
\mu_Z(T) &= \int_0^T\int_0^\infty \beta x e^{-(\beta+\delta)s}f_X(x)c_\theta\left(F_X(x),F_W(s)\right)dxds + \beta\int_0^T e^{-(\beta+\delta)s}\mu_Z(T-s)ds \\
&= C(T) + \beta\int_0^T e^{-(\beta+\delta)s}\mu_Z(T-s)ds \\
&= C(T) + \beta\int_0^T e^{-(\beta+\delta)(T-s)}\mu_Z(s)ds
\end{aligned} \quad (2)$$

and:

$$\mu_Z^{(m)}(T) = \beta[\int_0^T \int_0^\infty x^m e^{-(\beta+m\delta)s} f_X(x) c_\theta(F_X(x), F_W(s))\,dxds$$
$$+ \sum_{1\le j<m} \binom{m}{j} \int_0^T x^{-(m-j)} e^{-[\beta+m\delta]s} c_\theta(F_X(x), F_W(s))\, f_X(x) \mu_Z^{(m-j)}(T-s)dxds]$$
$$+ \beta \int_0^T e^{-(\beta+m\delta)s} \mu_Z^{(m)}(T-s)ds$$
$$= C^{(m)}(T) + \beta \int_0^T e^{-(\beta+m\delta)s} \mu_Z^{(m)}(T-s)ds$$
$$= C^{(m)}(T) + \beta \int_0^T e^{-(\beta+m\delta)(T-s)} \mu_Z^{(m)}(s)ds$$

(3)

where

$$C^{(m)}(T) = \beta[\int_0^T \int_0^\infty x^m e^{-(\beta+m\delta)s} f_X(x) c_\theta(F_X(x), F_W(s))\,dxds$$
$$+ \sum_{1\le j<m} \binom{m}{j} \int_0^T x^{-(m-j)} e^{-[\beta+m\delta]s} c_\theta(F_X(x), F_W(s))\, f_X(x) \mu_Z^{(m-j)}(T-s)dxds]$$

(4)

for m=2,3,⋯.

2.2. Copula Used

We are interested to calculate the first, second and *m*th moment of aggregate discounted claims under three copulas: the FGM copula, the Gaussian copula and the Gumbel copula. Their pdfs are given by:

$$c_\theta^F(F_X(x), F_W(s)) = 1 + \theta(1 - 2F_X(x))(1 - 2F_W(s))$$

(5)

$$c_\theta^G(F_X(x), F_W(s)) = \frac{1}{\sqrt{(1-\theta^2)}} e^{-\frac{\theta(2\Phi^{-1}(F_X(x))\Phi^{-1}(F_W(s)) - \theta(\Phi^{-1}(F_X(x))^2 + \Phi^{-1}(F_W(s))^2))}{2(\theta^2-1)}}$$

(6)

$$c_\theta^M(F_X(x), F_W(s)) = \frac{(-\ln u)^\theta}{-u\ln u}\frac{(-\ln v)^\theta}{-v\ln v}\frac{\sqrt[\theta]{(-\ln u)^\theta + (-\ln v)^\theta}}{[(-\ln u)^\theta + (-\ln v)^\theta]^2} \times \frac{\sqrt[\theta]{(-\ln u)^\theta + (-\ln v)^\theta} + \theta - 1}{e^{\sqrt[\theta]{(-\ln u)^\theta + (-\ln v)^\theta}}} \tag{7}$$

The FGM copula is used in this study due to its simplicity and analytical tractability. It is also used to verify our numerical results in Section 4.1 with [6]. The well-known elliptical family member, the Gaussian copula, is chosen as, to the best of our knowledge, the effect of elliptical copula in terms of the dependence between claim size and inter-claim time have not been explored extensively. The Gumbel copula is also chosen, since it could be adopted by an insurance company, who assumes risks with extreme magnitude, having the tendency to occur together, as pointed out by De Matteis in [21]. Many standard statistical texts offer illustrations of copula scatter plots with various dependence structure, for which we refer to [22,23,24].

3. LINEAR INTEGRAL EQUATIONS

The most general form of linear integral equation (IE) is given by:

$$h(T)\Psi(T) = g(T) + \int_a^{b(T)} K(T,s)\Psi(s)ds \tag{8}$$

where Ψ(T) is the solution to the IE that we need to obtain, g(T) is a given function and K(T,s) is the kernel for the IE. Equation (8) can be a homogeneous/non-homogenous, Volterra/Fredholm IE of the 1st/the 2nd kind, for which readers are referred to the conditions given in Section 2.1 of [18]. Linear IE can be solved either numerically using methods, such as the Runge–Kutta and collocation methods (see, e.g., [25,26]), or solved explicitly, such as by obtaining its Neumann series via the Picard method of successive approximations or using the Laplace transform method.

3.1. Volterra IE of the 2nd Kind

If we have $g(T) \neq 0$, $h(T) = 1$, and $b(T) = T$, Equation (8) becomes:

$$\Psi(T) = g(T) + \int_a^T K(T,s)\Psi(s)ds \tag{9}$$

which is a non-homogeneous Volterra integral equation of the second kind. The Volterra IE is widely used in the areas of viscoelasticity and electromagnetic to compute the dynamics of materials that "contain" memory, other than being useful in renewal theory and demography (see, e.g., [16,27], as well as the references therein for a more rigorous treatment on Volterra integral equations).

We easily notice that the moments provided by Equation (2) and Equation (3) take the form of Equation (9) and attempt to derive the explicit solution of the recursive expressions using Neumann series in the next subsection.

A unique and continuous solution, Ψ(T), is obtainable if we have a combination of a continuous kernel, K(T,s), in the region a≤s≤T≤b(T) with a function, g(T), that is continuous in the region a≤T≤b(T), even though it is not a requirement for the kernel function, K(T,s), to be continuous (see page 1 of [28] and page 5 of [20]). For the case of a discontinuous kernel function, we need to check if K(T,s) fulfills the three regularity conditions set on page 3 of [27], and, hence is an L2L2-function.

In the case of the first and second moments, μZ(T) and μ(2)Z(T), the function, g(T), is represented by the following equations, respectively:

$$\int_0^T \int_0^\infty e^{-(\beta+\delta)s} x f_X(x) c_\theta \left(F_X(x), F_W(s)\right) dx ds \tag{10}$$

$$\int_0^T \int_0^\infty e^{-(\beta+2\delta)s} x^2 f_X(x) c_\theta \left(F_X(x), F_W(s)\right) dx ds$$

$$+2 \int_0^T \int_0^\infty e^{-(\beta+2\delta)s} x f_X(x) c_\theta \left(F_X(x), F_W(s)\right) \mu_Z(T-s) dx ds$$

(11)

where $F_W(s)=1-e^{-\beta s}$ is the inter-claim time cdf.

As *X* and *W* are continuous r.v.'s, and by corollary 2.2.6 of [23] on copula continuity, g(T) is the continuous function for $s \in [0, T]$ and $x \in [0, \infty]$, since it is the sum and product of continuous functions. The kernel function is also continuous, as it is an exponential function given by:

$$K(T, s) = e^{-(\beta+m\delta)(T-s)} \tag{12}$$

Additionally, it is a bounded function in the square $\Pi = \{(T, s) : a \leq T \leq b(T), a \leq s \leq T\}$.

3.2. Neumann Series

In this section, we will find the Neumann series of the Volterra IE assuming the exponentially distributed inter-claim arrival time and a general claim size with continuous pdf. To do so, we start with a proposition from Chapter 3 of [27], which used the Picard method of successive approximation.

Proposition 1. ***Neumann Series for a Volterra IE of the 2nd Kind***

For the Volterra IE of the 2nd kind, as in Equation (9), *where g(T) and K(T,s) are* L2L2-*functions, its Neumann series is given by:*

$$\begin{aligned}\Psi(T) &= g(T) + \sum_{n=1}^{\infty} \lambda^n \int_a^T K(T,s)\Psi(s)ds \\ &= g(T) + \lambda \int_a^T \sum_{n=1}^{\infty} \lambda^{n-1} K_n(T,s) g(s)ds \\ &= g(T) + \lambda \int_a^T \Gamma(T,s;\lambda) g(s)ds\end{aligned} \tag{13}$$

where $\Gamma(T,s;\lambda) = \sum_{n=1}^{\infty} \lambda^{n-1} K_n(T,s)$ *is the unique resolvent kernel and* $K_n(T,s)$ *is the n-th-iterated kernel function satisfying the recurrence formula:*

$$K_n(T,s) = \int_s^T K(T,u) K_{n-1}(u,s)du \tag{14}$$

with K1(T,s)=K(T,s).

In order to prove our theorem, it is necessary to find the resolvent kernel, which is obtained in the following lemma.

Lemma 2. *Consider the kernel function given by Equation* (12). *For* m=1,2,..., *its resolvent kernel is therefore given by:*

$$\Gamma(T,s;\lambda) = e^{-m\delta(T-s)} \tag{15}$$

Proof. Using Equation (14), we obtain $K_2(T,s), K_3(T,s), \cdots, K_{n+1}(T,s)$ starting from $K(T,s) = K_1(T,s) = e^{-(\beta+m\delta)(T-s)}$.Letting $K_2(T,s), K_3(T,s), \cdots, K_{n+1}(T,s)$ and since $(s-T)^n = [-(s-T)]^n$ for even *n*, the resolvent kernel is then obtained by summing up Km(T,s) as follows:

$$\begin{aligned}\Gamma(T, s; \lambda) &= e^{-(m\delta+\beta)(T-s)} \sum_{n=1}^{\infty} \frac{[-(s-T)\beta]^n}{n!} \\ &= e^{-(m\delta+\beta)(T-s)} e^{-\beta(T-s)} \\ &= e^{-m\delta(T-s)}\end{aligned}$$

Now, we can obtain the expression for the first and second moment, which is the main result of this article.

Theorem 3. *The explicit solution of the first two moments are given by:*

$$\mu_Z(T) = \int_0^T \int_0^\infty e^{-\delta s} x \mathfrak{L}_\theta(x, s) dx ds + \beta \int_0^T \int_0^s \int_0^\infty e^{-\delta(T-s+u)} x \mathfrak{L}_\theta(x, s) dx du ds \tag{16}$$

$$\begin{aligned}\mu_Z^{(2)}(T) = &\int_0^T \int_0^\infty e^{-2\delta s} x^2 \mathfrak{L}_\theta(F_X(x), F_W(s)) dx ds \\ &+ 2 \int_0^T \int_0^\infty \int_0^{T-s} \int_0^\infty e^{-2\delta s - \delta\tau} x h \mathfrak{L}_\theta(F_X(x), F_W(s)) \mathfrak{L}_\theta(F_X(h), F_W(\tau)) dh d\tau dx ds \\ &+ 2\beta \int_0^T \int_0^\infty \int_0^{T-s} \int_0^\infty e^{-\delta(T+s-\tau+u)} x h \mathfrak{L}_\theta(F_X(x), F_W(s)) \mathfrak{L}_\theta(F_X(h), F_W(u)) dh du d\tau dx ds \\ &+ \beta \int_0^T \int_0^s \int_0^\infty e^{-2\delta(T-s-\tau)} x^2 \mathfrak{L}_\theta(F_X(x), F_W(s)) dx ds \\ &+ 2\beta \int_0^T \int_0^{s-\tau} \int_0^\infty \int_0^s \int_0^\infty e^{-2\delta(T-s-\tau)} x h \mathfrak{L}_\theta(F_X(x), F_W(\tau)) \mathfrak{L}_\theta(F_X(h), F_W(y)) dh dy dx d\tau dx ds \\ &+ 2\beta^2 \int_0^T \int_0^s \int_0^\infty \int_0^{s-\tau} \int_0^y \int_0^\infty e^{-\delta(2T+u-y-s+\tau)} x h \mathfrak{L}_\theta(F_X(x), F_W(\tau)) \\ &\times \mathfrak{L}_\theta(F_X(h), F_W(u)) dh du dy dx d\tau dx ds\end{aligned} \tag{17}$$

where $\mathfrak{L}_\theta(F_X(x), F_W(u)) = e^{-\beta u} f_X(x) c_\theta(F_X(x), F_W(u))$ *with* $F_W(u) = 1 - e^{-\beta u}$.

Proof. Applying Proposition 1 and Lemma 2 to Equation (3) with m=1,2, the results follow.

Section 4 will numerically illustrate the computation of the first and second moment under three copulas, assuming that the claim sizes are exponentially distributed, *i.e.*, X~exp(α). We do not proceed to obtain the closed form solution of the Neumann series expression for the higher moments, as it is tedious and time consuming. However, they are obtainable using the results provided in this section.

4. NUMERICAL ILLUSTRATION

We now present numerical illustration of the Neumann series expression for the first two moments. We start our discussion by presenting the scatter plots of each copula in Figure 1, Figure 2 and Figure 3, where the marginals are exponential distribution, which is in line with the assumptions used in the numerical computations of the moments in this section. All computations were done using Mathematica.

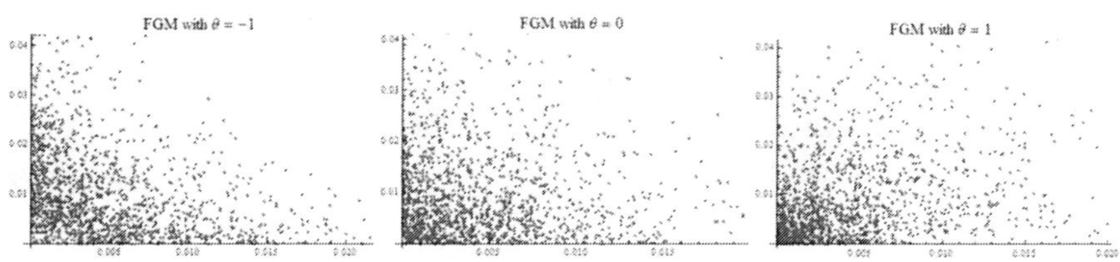

Figure 1. Farlie–Gumbel–Morgenstern (FGM) copula with exponential margins and dependence parameters −1, zero, one.

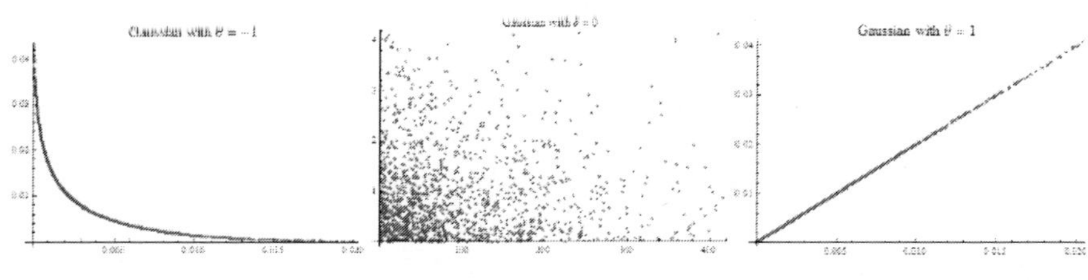

Figure 2. Gaussian copula with exponential margins and dependence parameters −1, zero, one.

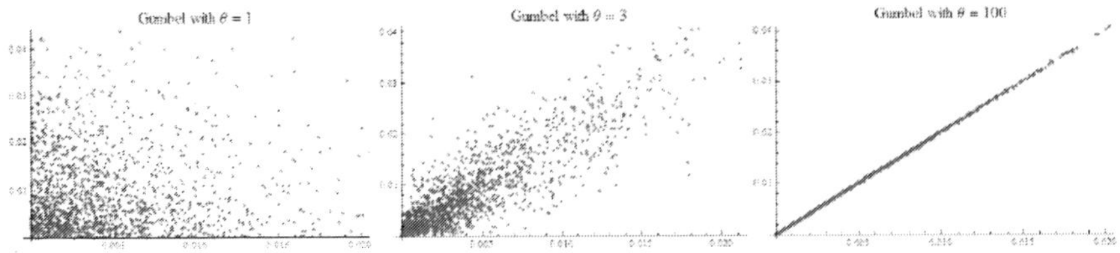

Figure 3. Gumbel copula with exponential margins and dependence parameters one, three, 100.

4.1. Numerical Accuracy of Neumann Series Expression for Moments

Recall that Equation (16) has at most triple integration involved, while Equation (17) has up to sextuple numerical integration. This implies that the computation of Equation (17) is expected to be close to the solution by [6], due to numerical approximation error, and the values would vary according to selected software packages. To evaluate the performance of the main results, we compare the numerical values returned by our Neumann series (under the column Neumann ofTable 1), using the FGM copula, with the numerical values given by the closed form solution in [6] (under the column BCLM of Table 1).

Table 1. Moment verification: the case of the FGM copula. Abs. Dev., absolute deviation; Rel. Dev., relative deviation.

Moment	θ	BCLM	Neumann	Abs. Dev.	Rel. Dev.
$\mu_Z(5)$	-0.9	0.475231	0.475231	0	0
	0	0.453173	0.453173	0	0
	0.9	0.431115	0.431115	0	0
$\mu_Z^{(2)}(5)$	-0.9	0.332023	0.329774	0.002249	0.006774
	0	0.287786	0.287786	0	0
	0.9	0.245457	0.247706	0.002249	0.009163

The values in Table 1 were computed using an example of δ=0.04, α=10, β=, T=5 and θ=−0.9,0,0.9. The absolute deviation (Abs. Dev.) figures are obtained by taking the difference between the two columns, BCLM and Neumann (*i.e.*, the absolute value of the solution presented in [6] minus the Neumann series expression), whereas the relative deviation (Rel. Dev.) figures are calculated as $\frac{Abs.Dev.}{BCLM}$.

Our calculations showed that the Neumann series expression for the first moment gives the same value as the closed form solution presented in [6]. On the other hand, the second moment gives a slightly different value at θ=−0.9 and 0.9, when the r.v.'s, *X* and *W*, are highly dependent. After a close scrutiny of the programming messages, we noticed that this is caused by numerical approximation errors of the quadruple, quintuple and sextuple integrations that are not present in the calculation of the first moment. To improve the accuracy of the Neumann series expression for higher order moments, the reader can use other software packages or use Monte Carlo simulation.

4.2. Moments of the Aggregate Discounted Claims

Setting δ=0.04, α=0.01and β=1 for the case of exponential claim inter-arrival time and exponential claim sizes, respectively, we show the values of moments of the aggregate discounted claims for each copula used in this study. We present the values of the first and second moments of the compound distribution, *i.e.*, $\mu_Z(5)$ and $\mu_Z^{(2)}(5)$, as well as the variance under each copula inTable 2, Table 3 and Table 4. The term "spread", which is defined as the difference between the values returned by θ at both ends, *i.e.*, $\theta_{-0.95} - \theta_{0.95}$ for FGM and Gaussian, and $\theta_1 - \theta_{100}$ for Gumbel, are also shown in Table 2 and Table 3.

Table 2. Values of μZ(5) for various copula.

θ	FGM	θ	Gaussian	θ	Gumbel
−0.95	455.543	−0.95	513.470	1	453.173
−0.9	455.419	−0.9	511.887	5	360.864
−0.5	454.421	−0.5	488.903	15	267.995
0	453.173	0	453.173	30	148.317
0.5	451.926	0.5	409.481	50	104.457
0.9	450.928	0.9	368.612	75	21.486
0.95	450.803	0.95	363.124	100	10.712
Spread	4.74		150.346		431.687

Our calculations showed that all copula exhibit decreasing values as θ increases, in line with [11]. Intuitively, a negative dependence structure represented by the pair of short inter-claim waiting time (or frequent claim occurrence within a given time period) with huge claim size will only prompt the insurer to charge a higher premium as opposed to the positive dependence structure.

As we have expected, the values of moments do not vary much across θ when calculated under the FGM copula, as opposed to the Gaussian and Gumbel copulas. Being an extreme copula, values of the first moment calculated under the Gumbel copula also showed the widest spread of the first moment.

Table 5 shows the values of the first moment as a function of α and β, respectively, for which we use the Gaussian copula at θ=−0.9. It shows that increasing the inter-claim waiting time parameter, β, results in increasing the mean value of the aggregate discounted claims, and *vice versa* in the case of the claim size parameter. Given an average value of inter-claim arrival time, β, the mean of aggregate discounted claims gets lower as we have a lower average claim size, given by 1α1α. On the other hand, given an average value of the claim size, the mean of the aggregate discounted claims gets bigger as the inter-claim arrival time gets shorter, which implies more frequent claim occurrences. This scenario is illustrated in Figure 4 for θ=−0.9 (left hand side of the diagram), as well as θ=0 (right hand side of the diagram).

Table 3. Values of $\mu_Z^{(2)}(5)$ for various copula.

θ	FGM	θ	Gaussian	θ	Gumbel
−0.95	336,551.170	−0.95	409,852.140	1	287,784.972
−0.9	332,022.549	−0.9	405,315.216	3	148,590.220
−0.5	312,126.218	−0.5	357,029.617	5	139,437.15
0	287,785.862	0	287,785.862	40	21,804.385
0.5	264,034.461	0.5	212,119.492	75	1,088.013
0.9	245,457.386	0.9	149,988.657	80	229.608
0.95	241,329.490	0.95	136,249.086	100	178.443
Spread	95,221.68		273,603.054		287,606.529

Table 4. Values of Var(5) for various copula.

θ	FGM	θ	Gaussian	θ	Gumbel
−0.95	128,940.627	−0.95	146,200.971	1	82,420.094
−0.9	124,616.083	−0.9	143,286.915	3	13,837.353
−0.5	105,627.773	0.5	118,003.474	5	9,214.320
0	82,420.094	0.	82,420.094	40	5,597.566
0.5	59,797.351	0.5	44,444.803	75	626.365
0.9	42,121.325	0.9	14,113.850	80	112.770
0.95	38,106.145	0.95	4,390.216	100	61.436

Table 5. Values of μZ(5) under the Gaussian copula at θ=−0.9.

$\beta = 1$	$\mu_Z(5)$	$\alpha = 1$	$\mu_Z(5)$
$\alpha = 0.01$	511.887741	$\beta = 0.01$	0.165673
$\alpha = 0.1$	51.188786	$\beta = 0.1$	0.884887
$\alpha = 1$	5.118887	$\beta = 1$	5.118887
$\alpha = 10$	0.511838	$\beta = 10$	45.911776
$\alpha = 15$	0.341257	$\beta = 15$	67.571651

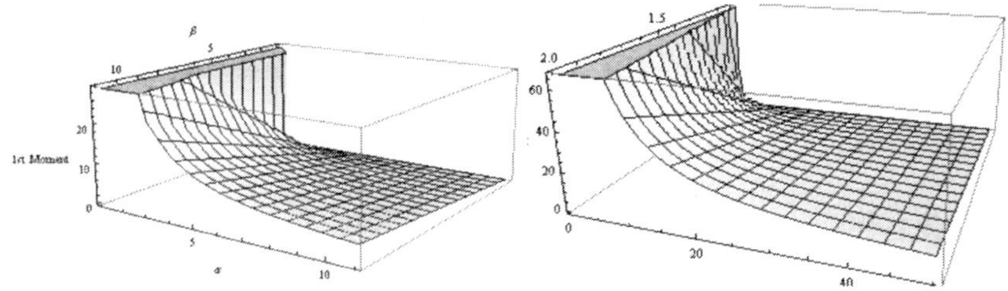

Figure 4. Sensitivity of the first moment under the Gaussian copula at θ=0 and θ=−0.9 with respect to claim size and inter-claim time averages.

4.3. Premium Calculation under FGM, Gaussian and Gumbel copulas

We now compute the loaded premium related to the risk of an insurance portfolio represented by Z(T)Z(T), where the dependence structure is captured by a copula. For that purpose, the first two moments will be useful in the premium calculation based on the expected value principle, the variance principle, as well as the standard deviation (SD) premium principle, as the following:

$$\Pi(T) = \mathbf{E}[Z(T)] + \kappa\mathbf{E}[Z(T)]$$

$$\Pi(T) = \mathbf{E}[Z(T)] + \kappa\mathbf{V}ar[Z(T)]$$

$$\Pi(T) = \mathbf{E}[Z(T)] + \kappa\sqrt{\mathbf{V}ar[Z(T)]}$$

Table 6 exhibits the loaded premium according to the SD principle under the three copulas considered, with κ=0.1, while Figure 5 and Figure 6 illustrate the range of premiums under the copulas studied according to the SD premium principle.

Table 6. Loaded premium according to the SD principle under various copulas.

θ	FGM	θ	Gaussian	θ	Gumbel
−0.95	491.55	−0.95	551.71	1	481.88
−0.9	490.72	−0.9	549.74	3	378.85
−0.5	486.92	−0.5	523.25	5	370.46
0	481.88	0	481.88	40	134.79
0.5	476.38	0.5	430.56	75	23.99
0.9	471.45	0.9	380.49	80	11.87
0.95	470.32	0.95	369.75	100	11.60
Spread	21.23		181.96		470.28

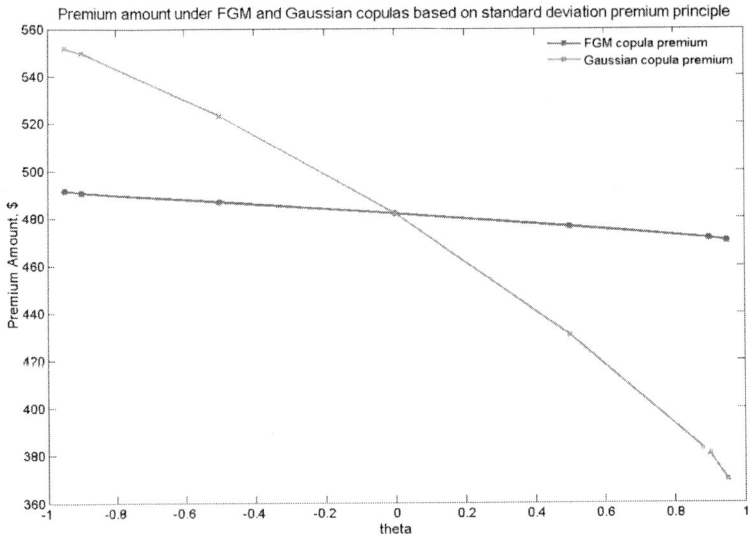

Figure 5. The loaded premium under FGM and Gaussian copulas based on the SD premium principle.

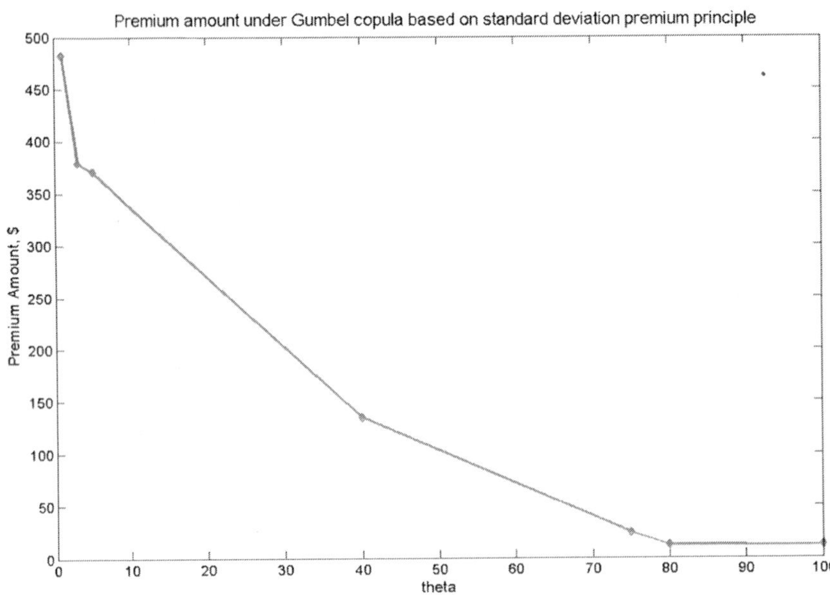

Figure 6. The loaded premium under the Gumbel copula based on the SD premium principle.

5. CONCLUSIONS

In this paper, we utilized copulas to capture the dependence structure between the inter-claim arrival time and claim sizes in classical actuarial risk theory. To do so, we represented the expression for the m−thm-th order moment proposed in [6,12] in the form of the Volterra integral equation (VIE) of the second kind, which is widely used in renewal theory, demographics, electromagnetism and viscoelasticity.

We derived the Neumann series expression for this recursive equation using the Picard method of successive approximations, based on which we computed the first two moments of the aggregate discounted claims. For the dependence structure between the inter-claim arrival time and claim sizes, we used a Farlie–Gumbel–Morgenstern copula, a Gaussian copula and a Gumbel copula with exponential marginal distributions. We showed the values of moments of the aggregate discounted claims, as well as the loaded premium for each copula used in this study.

It would be of interest to derive the expression for Equation (2) and Equation (3) using other joint pdfs between X and W. Other copulas with different claim size distributions for X may be considered in the proposed approach, which we leave for further research. We can also consider the Monte Carlo simulation, as well as other numerical methods to solve the VIE (such as Runge–Kutta and the collocation methods), as the next objective of further research to deal with the computation of higher moments.

ACKNOWLEDGMENTS

The correspondent author is thankful to the Malaysian Ministry of Education and the National University of Malaysia for providing the educational fund to further PhD studies and for the opportunity to carry out this research. The authors also thank the anonymous referees for their comments.

AUTHOR CONTRIBUTIONS

Both authors contributed to all aspects of this work.

REFERENCES

1. A.M. Best Company. *Press Release Special Report: Regional Cat Losses Drive Asian Reinsurers to Focus on Profitability, Capital Strength*; A.M. Best Company: Hong Kong, China, 2012.

2. Munich Re. Review of Natural Catastrophes in 2011: Earthquake Result in Record Loss Year. Press Release. 2012. Available online: http://www.munichre.com/en/media_relations/press_releases/2012/2012_01_04_press_release.aspx (accessed on 21 October 2013).

3. Albrecher, H.; Boxma, O.J. A ruin model with dependence between claim sizes and claim intervals. *Insur. Math. Econ.* **2004**, *35*, 245–254.

4. Albrecher, H.; Teugels, J.L. Exponential behavior in the presence of dependence in risk theory. *J. Appl. Probab.* **2006**, *43*, 257–273.

5. Asimit, V.A.; Badescu, A.L. Extremes on the discounted aggregate claims in a time dependent risk model. *Scand. Actuar. J.* **2010**, *2*, 93–104.

6. Bargés, M.; Cossette, H.; Loisel, S.; Marceau, E. On the moments of aggregate discounted claims with dependence introduced by a FGM copula. *ASTIN Bull.* **2011**, *41*, 215–238.

7. Jang, J. Martingale approach for moments of discounted aggregate claims. *J. Risk Insur.* **2004**,*71*, 201–211.

8. Li, J.; Tang, Q.; Wu, R. Subexponential tails of discounted aggregate claims in a time-dependent renewal risk model. *Adv. Appl. Probab.* **2010**, *42*, 1126–1146.

9. Kim, B.; Kim, H.S. Moments of claims in a Markovian environment. *Insur. Math. Econ.* **2007**,*40*, 485–497.

10. Marri, F.; Furman, E. Pricing compound Poisson processes with the Farlie–Gumbel–Mogenstern dependence structure. *Insur. Math. Econ.* **2012**, *51*, 151–157.

11. Woo, J.-K.; Cheung, E.C.K. A note on discounted compound renewal sums under dependency. *Insur. Math. Econ.* **2013**, *52*, 170–179.

12. Léveillé, G.; Garrido, J. Recursive moments of compound renewal sums with discounted claims. *Scand. Actuar. J.* **2001**, *2*, 98–110.

13. Léveillé, G.; Garrido, J. Moments of compound renewal sums with discounted claims. *Insur. Math. Econ.* **2001**, *28*, 199–219.

14. Léveillé, G.; Garrido, J.; Wang, Y.F. Moment generating functions of compound renewal sums with discounted claims. *Scand. Actuar. J.* **2010**, *3*, 165–185.

15. Woo, J.-K. Some remarks on delayed renewal risk models. *ASTIN Bull.* **2010**, *40*, 199–219.

16. Bellman, R.E.; Cooke, K.L. *Differential-Difference Equation*; Academic Press: New York, NY, USA, 1963.

17. Burton, T.A. *Volterra Integral and Differential Equation*; Academic Press: New York, NY, USA, 1983.

18. Kotsireas, I. *A Survey on Solution Methods for Integral Equations*; UWO ORCCA Technical Reports; The Ontario Research Centre for Computer Algebra: London, ON, Canada, 2008; Volume 47.

19. Pruss, J. Evolutionary integral equations and applications. In *Springer Monographs in Mathematics*; Birkhauser: Basel, Switzerland, 1993; Volume 87.

20. Shestopalov, Y.V.; Smirnov, Y.G. *Lecture Notes on Integral Equations—Compendium*; Division for Engineering Science, Physics and Mathematics, Karlstad University: Karlstad, Sweden, 2002.

21. De Matteis, T. Fitting Copula to Data. MSc Thesis, Institute of Mathematics, University of Zurich, Zurich, Switzerland, 2001.

22. Denuit, M.; Dhaene, J.; Goovaerts, M.J.; Kaas, R. *Actuarial Theory for Dependent Risks: Meaures, Orders and Models*; Wiley: New York, NY, USA, 2005.

23. Nelsen, R.B. *An Introduction to Copulas*, 2nd ed.; Springer Series in Statistics; Springer: New York, NY, USA, 2006.

24. McNeil, A.J.; Frey, R.; Embrechts, P. *Quantitative Risk Management*; Princeton University Press: Princeton, NJ, USA, 2005.

25. Brunner, H. *Collocation Methods for Volterra Integral and Related Functional Differential Equations*; Cambridge University Press: Cambridge, UK, 2004.

26. Makroglou, A.; Konstantinides, D.G. Numerical solution of a system of two first order Volterra integro-differential equations arising in ultimate ruin theory. *HERMIS J.* **2006**, *7*, 123–143.

27. Kanwal, R.P. *Linear Integral Equations: Theory & Technique*, 2nd ed.; Modern Birkhauser Classic; Birkhauser: Boston, MA, USA, 2013.

28. Evans, G.C. Volterra's integral equation of the second kind, with discontinuous kernel.*Trans. Am. Math. Soc.* **1910**, *11*, 393–413.

Index